Fridtjof
Nansen
Spitsbergen
ARCTIC OCEAN
New Siberia Island
Greenland Sea
Novaya Zemlya
Kara Sea
North Cape
Barents Sea
CIRCLE
Iceland
Siberia
Yenisei River
Lena River
Dvina R.
URAL MTS.
Ob River
EUROPE
ASIA
North Sea
Baltic Sea
Volga R.
Dnieper R.
Irtish
Lake Baikal
Sea of Okhotsk
Sakhalin
Amur River
ALPS
Danube R.
Black Sea
Caspian Sea
Aral Sea
Lake Balkhash
Amu Darya R.
Mediterranean Sea
HIMALAYA MTS.
Hwang-ho (Yellow R.)
Yangtze Kiang
Sea of Japan
JAPAN
Madeira
Canary Islands
Euphrates R.
Suez Canal
Indus R.
Brahmaputra R.
Ganges R.
Salween R.
Mekong R.
Sahara Desert
L. Chad
Nile R.
Red Sea
Arabian Sea
Bay of Bengal
Irrawaddy River
South China Sea
Philippine Islands
PACIFIC OCEAN
Niger R.
AFRICA
Ceylon
Gulf of Guinea
L. Victoria
Sumatra
Borneo
Celebes
New Guinea
•Ascension I.
Congo River
L. Tanganyika
Lake Nyasa
INDIAN OCEAN
Java
ATLANTIC OCEAN
Saint Helena
Zambesi R.
Madagascar
AUSTRALIA
Orange R.
Cape of Good Hope
Murray R.
Tasmania
ANTARCTIC OCEAN
ANTARCTICA

Lands and Peoples

THE WORLD IN COLOR

VOLUME VI

THE GROLIER SOCIETY

NEW YORK TORONTO

Copyright © 1956, 1955, 1954, 1953, 1952, 1951, 1949, 1948, 1946, 1943, 1941, 1940, 1938

by THE GROLIER SOCIETY INC.

Library of Congress Catalog Card Number: 56–5009

Copyright 1932, 1930, 1929 by THE GROLIER SOCIETY

Copyright © 1956, 1955, 1954, 1953, 1952, 1951 by THE GROLIER SOCIETY OF CANADA LIMITED

Volume VI
TABLE OF CONTENTS

	PAGE
THE FRIENDLY NORTH—*Northern Canada and the Polar Regions*	4
AMERICAN INDIANS—*The Tribes of Canada and the United States* 8 Pages in Full Color	17
CANADA, MONARCH OF THE NORTH—*The Development of a Colony into a Nation*	33
THE ATLANTIC PROVINCES—*Lands Rich in the Traditions of the Sea* 8 Pages in Full Color	43
QUEBEC, ONCE NEW FRANCE—*The Largest Province of Canada* 8 Pages in Full Color	65
ONTARIO, FORMERLY UPPER CANADA—*The Central Province of the Dominion* 8 Pages in Full Color	81
THE WESTERN PROVINCES—*British Columbia and the Prairie Provinces* 8 Pages in Full Color	97
CANADIAN CITIES—*Great Variety and Striking Contrasts*	121
THE ROOF OF NORTH AMERICA—*The Yukon and the Northwest Territories*	144
THE UNITED STATES—*How the Republic Spanned a Continent* 4 Pages in Full Color	153
FROM MAINE TO MARYLAND—*Town and Country in the Northeast* 8 Pages in Full Color	201
FROM VIRGINIA TO TEXAS—*The Story of the Southern States* 8 Pages in Full Color	225
STATES OF LAKE AND PLAIN—*The North Central States* 8 Pages in Full Color	249
THE STATES TOWARD THE SUNSET—*Mountain and Pacific States* 12 Pages in Full Color	272
LANDS OF TREASURE AND ROMANCE—*Territories of Hawaii and Alaska* 4 Pages in Full Color	305
THE CITIES OF THE UNITED STATES—*Its Centers of Culture and Industry*	317
BEAUTY, WONDER, WISE HUSBANDRY—*National Parks and Forests* 8 Pages in Full Color	376
UNSPOILED WILDERNESS LANDS—*National Parks and Reserves of Canada*	401
COLOR PLATES IN VOLUME VI	432

HUGE ICEBERG SEEN FROM A SAILBOAT'S DECK OFF BAFFIN LAND

Wide World Photos

THE FRIENDLY NORTH
Northern Canada and the Polar Regions

For a thousand years men have sought the secrets of the Arctic regions. The desire to find a Northwest Passage over the top of North America led to many expeditions, of which the most famous was that of Sir John Franklin (1845-48), who perished with all his men. During the latter half of the nineteenth century there were dozens of Arctic expeditions. A United States naval officer, R. E. Peary, spent several seasons in the Arctic regions and in 1909 announced that he and five companions had reached the Pole April 6, 1909, using dog sledges. On May 9, 1926, another United States naval officer, Richard E. Byrd, accompanied by Floyd Bennett, flew over the Pole in the airplane, Josephine Ford. The first crossing of the polar sea by airship was made by Amundsen, Ellsworth and Nobile in the Norge, in 1926; while the first crossing by airplane was by Wilkins and Eielson in 1928. Meanwhile others had been engaged in a varied program of scientific exploration. Perhaps the most successful was Vilhjalmur Stefansson who headed three expeditions, exploring and studying much of hitherto unknown Arctic. In the story that follows we tell you something of the region and its inhabitants.

BY a sort of impromptu legal mathematics, Canada has defined her sub-Arctic possessions. South of parallel 60 she speaks of provinces, but north of that degree of latitude she calls them territories. Like most rules, however, this one has an exception, for a small triangle of Quebec protrudes north of the boundary.

The lands north of 60° are nearly half of Canada. When we remember that Canada is larger than the forty-eight states of the United States we are prepared to find within these northern domains many conditions, several climates, and at least three conspicuously distinct types of people. The vegetation and the animals differ correspondingly.

Yukon Territory is mostly mountainous and forested. The high range along the southwest corner prevents the inland climate from being much affected by the Pacific, and there is only a little panhandle stretching north which is materially affected by the sea climate of the Arctic. Most of the Yukon has, therefore, a continental climate. The summers are hot, with temperatures ranging upward to 100° in the shade; the winters are cold, with the alcohol thermometers falling toward 80° below zero.

The Northwest Territories comprise all the remaining land of Canada including the adjacent islands not included in any province. The forests and prairies of the Northwest Territories, east of the Yukon, are determined by the trend of midsummer sea winds. Therefore, the boundary between treeland and grassland runs more or less northwestward from a point a little north of Churchill, on Hudson Bay, to the Polar Sea near the foot of the Cape Parry Peninsula, straight south from Banks Island. South or southwest of this line we have roughly the continental or Yukon climate, with intense summer heat and intense winter cold. Northeast of the line the winters are less cold and the summers less warm. The islands north of Canada are necessarily prairie, for the same reason of midsummer sea winds.

Greenland is, by the current Danish estimates, 84 per cent ice-covered. This is because that territory is mountainous and has a heavy precipitation from the surrounding waters. No Canadian island is so high and none of them are therefore ice-capped, but there are considerable glaciers in three—Ellesmere, Heiberg and North Devon. There are some glaciers in Baffin Island but it is doubtful whether they are as large as those of British Columbia, although probably larger than the ones so familiar to tourists in Switzerland. There may be a glacier

MAP OF THE ARCTIC REGIONS

This map shows the vast areas that lie beyond the Arctic Circle. Note that much of this forbidding region is covered by the icy waters of the Arctic Ocean. In the corresponding region of the Antarctic, however, we find a great continent, called Antarctica. On the map we show the routes of five important expeditions of the twentieth century. The American naval officer Peary and his men made their dash to the North Pole in dog-sleds (1909). Byrd, another American naval officer, used the aeroplane; so did the Russians Schmidt and Vodopyanoff (1937). Dirigibles were used by the Amundsen-Ellsworth-Nobile expedition of 1926 and the Nobile expedition of 1928.

on Meighen Island. Apart from this there is far less permanent snow in the Canadian islands than there is in Switzerland and Austria. The Canadian mainland north of the Arctic Circle has no permanent snow. This is because mountains are absent and the winter snowfall is light. Portions of such states as New York and Michigan have an annual snowfall from two to five times as great as the average for the Canadian Arctic.

The old popular idea of Canada, Our Lady of the Snows, as part of the "frozen north," sometimes even exaggerated into "barren ground," has long been exploded, yet the editor of the Northwestern Miller, perhaps the leading journal of that industry in the world, has said that he well remembers the time when the millers of the Twin Cities of Minnesota had little confidence in the permanent wheat-producing power of northern North Dakota,

THE FRIENDLY NORTH

and believed that no serious competition would ever come from the British prairies to the north. But he adds that now Winnipeg alone handles more wheat than the three largest wheat markets of the United States put together.

A more remarkable or at least a quicker change of opinion has taken place within the Canadian Prairie Provinces themselves. No one in Alberta is entitled to membership in an Old Settlers Association who does not remember the time when Edmonton, just south of the centre of that province, was considered to be on the northern fringe of wheat production. Now it is common talk, even in Calgary, Edmonton's more southerly rival, that the northern half of the province will eventually produce more wheat than the southern half. The prize for the world's best wheat has already been won several times by the northern half of Alberta Province.

What surprises even Canadians is that summer frosts which injure wheat are actually more numerous south of the middle of Alberta than north of it. The point is that the maximum heat of the summer noonday, around 95° or 98° in the shade, remains about the same as you go north from the equator, while the sun shines more and more hours per day. The night is therefore shorter and the earth and air have less time to cool off between the last warming of the sunset and the first warming of the sunrise. When you get so far north that the midsummer nights are nearly gone the night frosts of midsummer are completely gone.

For this reason it is possible that farming will become a major occupation, though it probably will never surpass mining in importance. The development of mines has progressed rapidly and great quantities of gold and oil are now being produced. Furthermore, the area around Great Bear Lake is one of the world's chief

STANDARD OIL CO. (N. J.)

WARMLY CLAD INDIAN CHILDREN go to school in Aklavik, a trading village on the delta of the Mackenzie River. Trapping fur-bearing animals is a major industry in this region.

THE FRIENDLY NORTH

Courtesy Canadian Pacific Railway
A TYPICAL MEMBER OF THE ROYAL CANADIAN MOUNTED POLICE

All members are trained as horsemen, though the force also uses autos and airplanes.

sources of radium and uranium ores.

But these new industries require people, and as more and more people move into the area, agriculture becomes increasingly important. The government agricultural station near Whitehorse produced fine crops of wheat and rye as early as 1948.

It is now believed that the land can produce more than enough to feed its growing population and it is possible that the wheat empire of Canada may stretch north into the subarctic Northwest Territories.

The books used to say that in the Northwest Territories of Canada the ground is always frozen and "the vegetation therefore poor or absent." But the ground frost really produces an effect just opposite to the one we had expected. A hundred miles south of the Arctic Circle one smart kick with a booted toe reveals in August ground underneath the warm sod perpetually frozen as hard as granite, but a hundred miles north of the Circle you still find growing upon this icy concrete trees of white spruce measured by the Forestry Commissioner of Canada at over seventy feet.

However, the most definite friendliness of the ground-ice for vegetation is seen on the prairies north of the forest. For the frost nullifies that yearly variation in rainfall which in most lands is considerable and in some extreme. Plants would die in Australia under rains as scant as those of the Arctic. But in northern Canada when a dry season produces increased heating of the soil, there follows a melting of a little of the ice below, so that all the plants have to do to quench their thirst abundantly is to reach with their roots an inch or two deeper than they usually do. There is accordingly neither in northern Canada nor in northern Siberia any appreciable variation in the productivity of the grasslands between the dry and wet years.

At Fort Yukon, Alaska, north of the Arctic Circle, a tested United States Weather Bureau thermometer has recorded 100° in the shade six feet above eternal frost. One of those six feet is soil which so completely imprisons the chill of the ground that the heat of the air just above it is that of the humid tropics, whether judged by an instrument or by the nerves of the people who swelter in it.

There is intense winter cold in the places of the greatest sub-Arctic summer heat. The spot that has 95° in the shade in July may have —75° in winter. However, the coldest places, both in Canada and Siberia, are found within the forest. Out on the prairie, which may be called "barren grounds" if you want to be terrifying or romantic, you seldom come within ten or fifteen degrees of weather as cold as that of the forest. This seems to be because the prairie condition is produced by the chill ocean winds of summer. These same ocean winds are correspondingly warmer in winter. Therefore you find that the minimum temperatures get less and less severe as you go north through Canada in January toward the Arctic Sea. Seventy-five below zero is recorded frequently in the Canadian forest, but fifty-five below has never yet been recorded on the north coast of Canada. Neither is there any weather bureau or other probably reliable record as low as —55° on any of the islands to the north of Canada.

These islands, however, are never extremely hot in summer. You must get far from the ocean to have great heat, and by the nature of things you cannot do this on an island. Similarly by the nature of things you will have the greater heat inland the larger the island. Tempera- Siberia. But the important thing is that when you come to the end of the forest you only come to the beginning of the prairie. You may disguise that fact, if you like, by calling it "barren ground" in Canada or "tundra" in Siberia, but that is mostly quibbling. True enough, there is a difference between the Arctic and the Temperate Zone prairie, just as there is a difference between the prairies

STANDARD OIL CO. (N. J.)

WINTER TWILIGHT IN NORMAN WELLS, IN NORTHWESTERN CANADA
The town is on the Mackenzie River almost at the Arctic Circle. Originally it was a trading post but it has become important in recent years as the center of a rich oil field.

tures around 85° in the shade will be found in the centre of Victoria Island, 300 to 400 miles north of the Arctic Circle. None so high are probably found in Baffin or Ellesmere Islands. This is not because they are more northerly, but because they are both higher and swept by more persistent sea winds.

The forests of spruce extend more than a hundred miles north of the Arctic Circle in some of the river valleys of Canada and more than twice that far in of Montana and Brazil. However, if you want to convey the idea that to the casual eye there is much similarity between treeless but well-watered grasslands in every zone, then the best common word is prairie.

There are some districts in the Arctic, no doubt, where mosses and lichens prevail above flowering plants both in number of species and in tonnage per acre. But in the Arctic as a whole there are 700 species of flowering plants against 500

ESKIMOS POSING FOR A MOTION PICTURE
Pulling up on a cake of sea ice a bear they have killed.

Photographs by Ewing Galloway
ESKIMO GIRLS STARTING OFF FOR A DUCK HUNT
Inset: The results.

The implication that the girl secured her ducks with bow and arrow shows the imagination of the movie director. Before guns were introduced it was the men who hunted ducks, not the women, and they used a kind of bolas, not a bow.

COPPER ESKIMO CARIBOU HUNTERS IN 1915

They used bows then but have rifles now. One of these men is wearing snow goggles of wood with narrow slits, the other has them raised on his forehead. The Copper Eskimos are so called because most of their weapons were of hammered and ground native copper. They live around Coronation Gulf and Victoria Island.

of mosses and lichens combined; by tonnage there is at least ten times as much flowering vegetation in the Arctic as non-flowering. Texas claims only 500 species of native flowering plants.

Where there are flowers there are certain to be insects. Peary saw a bumble-bee out over the ocean half a mile north of the most northerly land on earth. De Long's men caught a live butterfly on the floating sea-ice, and this was 700 miles north of the Arctic Circle and 10 or 20 miles from the nearest island.

Mosquitoes are the great plague and hardship in all inland parts of the sub-Arctic and on most Arctic islands. They get steadily more numerous as you go north through Canada from the United States boundary, until they are at their worst on or just south of the Arctic Circle. Then they get less as you continue north and are not serious any more 500 miles beyond the Circle. There is a similar northward decrease of many other insects.

The varieties of Arctic and sub-Arctic climate and conditions strike you particularly in relation to the people. Some Eskimos live in a forest but others have never been within several hundred miles of a tree. Most Eskimos live on or near a seacoast but there are some who have never been to the ocean. Fully half of the 35,000 or so Eskimos of the world live on seals mainly, but there are a few who have never tasted seal meat. Most Eskimos have still their native speech, but a few speak no language but English. Many in Greenland are familiar with Danish, some in northeastern Siberia know Russian, and so the complexity grows.

The first Eskimos came in contact with Europeans on the coast of Labrador about 900 years ago, and others on the coast of Greenland soon thereafter; but my second expedition in 1910 visited several hundred Eskimos who had never seen a white man until they met our party. It is probable that the last Eskimos saw their first European when the Rasmussen Expedition

Courtesy Department of Mines, Geological Survey

THE CARIBOU HUNTERS READY TO SHOOT

BUILDING A TEMPORARY SNOWHOUSE

The Eskimo mason quickly shapes and fits the snow blocks for a house in which to spend a day or two while out on a winter journey. A few thousand Eskimos live in snowhouses all winter, but a much larger number have never seen this kind of a dwelling.

THE SHELTER NEARS COMPLETION

Snow suggests cold to most people but to certain branches of the Eskimo people it is normal building material for a midwinter house in which he and his family are comfortable by day or by night.

GROUP OF COPPER ESKIMOS WITH THEIR SNOWHOUSES IN THE BACKGROUND
It is spring—early May. Some of the snow roofs have melted and caved in. They have been replaced by skin roofs spread over the snow walls.

came to them in 1923. Some Eskimos saw their first book or paper either in my own hands or Rasmussen's, but one of the oldest journals now published in the New World is as completely Eskimo as the Spectator is English or the Atlantic Monthly American, and has appeared every year since 1861.

Being the last people on the far edge of the earth, these northerners have been particular victims of our folklore and superstition. Apparently because it was a common European belief in ancient times that there were pigmies in the Far North, and also because cold is supposed to have a stunting effect, the Eskimos have been described until recently as a small or dwarfed people. They are more properly described as of medium size. Our idiom compels us to say "Eskimos and Indians" but the general scientific opinion is that they are merely one kind of Indian and should therefore be called Eskimo Indians, corresponding to Sioux Indians or Iroquois Indians.

There is a belief common even now that most Eskimos, or all of them, live in snowhouses in winter, but the fact is that snowhouses are about as local in the Arctic as adobe houses are in the United States. Europeans, wherever in the world you find their descendants, usually travel a great deal, see pictures and read books. Most Swedes, for instance, would know that there are adobe houses in New Mexico. But before the white man came

Courtesy Department of Mines, Geological Survey

A SUMMER CAMP OF THE COPPER ESKIMOS

NATIONAL FILM BOARD OF CANADA

NEXT WINTER'S FOOD for the dogs in the land of the long night. On Baffin Island, Eskimo Idlouk and his son, Oodlootituk, pile rocks on a cache of killed seals.

the Eskimos had no books, they traveled comparatively little, and some of them dwelt as far from others, when measured by the routes they had to travel, as Canada is from Brazil. Snowhouses have been seen by less than a third of the living Eskimo population of today. Most of them, however, know snowhouses pretty well through hearsay. A good many have seen movies made where they are found.

Eskimos are, generally speaking, a people of restless intelligence. Make it really clear to one inhabitant of a village that it is possible to set down a black mark on a white surface that means one sound and another black mark that means another sound. Show him that by twenty-five or thirty such marks, each different from the others, he can represent most or all of the important sounds of his own language. That is all you need to do. Come back a year or two later and you will find half the village reading and writing, with the knowledge of these skills already spreading to neighboring communities.

This, of course, applies only to writing the native tongue. American teachers in Alaska find the same difficulty teaching English to Eskimos that they would find in teaching Latin in Utah or Texas. But in a mining town, Eskimo children learn English in school about as rapidly as the whites, if given the same opportunity. That is because they also hear it in the street.

Arctic travelers usually agree that the Eskimos are the happiest people on earth. This could not be even half true if their lot were as hard as we used to suppose. It is curious how our books formerly told us, first, that life in the Arctic is necessarily a continuous hand-to-hand struggle with frost and famine, and second that the Eskimos have elaborate carvings in ivory and that their garments are frequently made up of thousands of separate pieces artfully sewn together into complicated designs. The truth is in the carvings and in the clothing; the false inference relates to the supposed desperate struggle for life. A woman who could sew for herself a warm coat in two days, if she made it from two whole caribou skins, will instead spend more than half of each day for two or three months in cutting up a great many skins into almost an infinite number of small pieces and then matching them together, eventually developing a coat that is neither so warm as the two-day garment would have been nor so durable. In like manner her husband spends whole extra days and half-days in carving the handle of a bag that would have served him as well if left plain.

The so-called civilized nations have in one city the contrast between stark poverty and surfeiting riches. Naturally there is within the vast Eskimo territories a similar contrast. There are indeed communities, and I have lived in some of them, where it is hard work to make both ends meet. I have seen this in Victoria Island, and in Iowa, with this difference, that the Iowa farmers never starve to death but the Eskimos sometimes do. That is primarily for transportation reasons.

We have, then, in the more or less far north, vast territories, thousands of miles of coastline, and great stretches of inland wood or prairie in which there live a varied but generally carefree and happy people known as Eskimos. The main thing that binds them together and makes them Eskimos is their common speech, one of the most difficult languages in the world for an outsider to learn.

It is said that a business man in a great city can get along if he has a ready command of from three to five thousand words, but an Eskimo cannot deal with his neighbors in less than ten to twelve thousand words, each colloquially at the tip of his tongue. It is not merely the size of the vocabulary. An English noun, for instance, has four forms—*man, man's, men, men's;* a Greek noun has nine forms; but an Eskimo noun has or can have more than a thousand forms, each different

EWING GALLOWAY

A PROUD INDIAN mother in the Yukon River country cradles her baby on her lap.

from any other and each with a precise meaning of its own. Their verbs are even more complicated than the nouns. Besides all that, you have to acquire a new way of thinking before you can speak such a tongue easily—a polysynthetic language.

But when you have learned a good polysynthetic language, you will not by choice use, say, English or German, for it is so much more flexible, precise and concise. Record, for instance, some Eskimo folktale as it is dictated to you by a storyteller. Then translate it idiomatically and you will find one page of Eskimo giving about two and a half pages of English.

THE FRIENDLY NORTH

You can say as much in one hour of Eskimo as in two and a half hours of English, and say it with more assurance that you have conveyed your real meaning.

Statements, such as those just made, about the Eskimo language sound curious to the layman, but they are ordinary to the student of languages. More than a generation ago, one of the greatest linguists that the United States ever produced, Whitney of Yale, said about primitive tongues in general what we have just said about Eskimo in particular. Had a people with the other gifts and the fortunate geographical and historical position of the English or the French, for instance, possessed at the same time a language like the Eskimo, they would in all probability have made even more rapid and substantial progress in civilization.

Some people get satisfaction from thinking how different other people are from themselves. The Eskimos have been a particular butt of this weakness, which goes to extreme lengths at times. There is, for instance, in Canada, a city where it is an important industry to build power schooners for sale to Eskimos. A citizen of this town has a photograph showing $200,000 worth of these schooners (each valued at $5,000) in one view. They are lighted electrically, and when winter comes and the boats are laid up the power plants are frequently taken out and used to light the houses. Some of these Eskimos have independent Delco lighting systems. Yet the schools of the city were until recently teaching the children that "the Eskimos" have no boats except of skin and no lights except seal oil lamps.

There are, true enough, Eskimos within 500 miles from the southern boundary of Canada who live in impoverished and primitive style, but there are also Eskimos farther north and farther from the great world centers who have electric lights, radios, and who use recording phonographs in their correspondence. The whole thing is very complicated, not open to any simple explanation. It is similar to the poverty one sometimes finds in mountainous regions, though wealthy cities may be near by. Such a strange contrast in living conditions can never be explained merely by distance north or south, or east or west.

YUKON AND NORTHWEST TERRITORIES: FACTS AND FIGURES

YUKON

A territory of northwestern Canada, bounded north by the Arctic Ocean, east by the District of Mackenzie, south by British Columbia and southwest and west by Alaska; area, 207,076 square miles (land area, 205,346 square miles); population (1951 census), including whites, Indians and Eskimos, 9,096. It is governed by a commissioner and a council of 3 elected members. Mining is the principal industry; gold, silver, lead and zinc are the leading metals produced. Lumbering is minor, but there are extensive stands of spruce, balsam, poplar, cottonwood and birch. Fur trading is important as a native industry. There are 58 miles of railway which link Whitehorse with Skagway, an Alaskan tidewater port. 570 miles of the Yukon River are navigable within the territory. There are nearly 3,000 miles of roads and trails, including 600 miles of the Alaska Highway. Telephone, telegraph, radio, postal and air services connect the towns with one another and with Alaska, the rest of Canada and the United States. In 1950 there were 11 schools with about 770 pupils. Population of chief towns: Whitehorse, 2,594; Dawson (capital), 783.

NORTHWEST TERRITORIES

Divided into the districts of Mackenzie, Keewatin and Franklin, it is bounded north by the Arctic Ocean, east by Baffin Bay, Davis and Hudson straits and Hudson Bay, south along the sixtieth parallel by the western provinces and west by Yukon Territory; area, 1,304,903 square miles (land area, 1,253,438 square miles), and population, 16,004. Governed by a commissioner and a council of 6—3 appointed and 3 elected. Mining, fishing and fur trading are the leading industries. Chief mineral products are uranium, gold, silver, crude oil and natural gas. Reindeer herding is encouraged. Motor roads mainly near posts and settlements; boat service on lakes and rivers. Airplanes are the principal means of travel. There are government and private radio stations. The territorial administration maintains 12 schools; private companies, 2; the Indian Affairs Branch of the Department of Citizenship and Immigration maintains 7; and several mission schools are assisted by government funds. Chief settlements are Yellowknife and Aklavik. Principal among the islands that make up Franklin District are Baffin, Victoria and Ellesmere.

AMERICAN INDIANS
The Tribes of Canada and the United States

Before the white man came the Indians lived close to nature—in forests, on plains, along seacoasts. Their customs, religious beliefs, their ways of securing food and shelter and clothing were closely related to the natural world. Gradually all this has changed, sadly in some ways and yet inevitably so. Today many Indians live in cities and towns and on farms just as other citizens of Canada and the United States do. Even on the reserves and reservations old ways are passing. More Indian boys and girls are going to school, some on to college, and their new ideas are a strong influence.

THE ancestors of today's Indians are thought to have come to America from Asia somewhere between ten and twenty-five thousand years ago. This does not mean that they were Chinese or Mongols, for those modern races had not yet developed. At the same time it is possible that the early Asiatic people from whom the Indians descended were also the ancestors of the modern yellow races and perhaps of some whites and blacks. The earliest arrivals came across what is now Bering Strait but may then have been dry land. Later, as men learned to build ships, some hardy adventurers may have crossed the Pacific.

Group by group the arrivals flooded in, ultimately populating the two Americas. They made the trails that have become our highways, they discovered every wild plant that was good for food and some of our present medicines and perfumes. Finally, in some warm country, perhaps South America, they began to sow the seeds of wild plants instead of wandering around to look for them. Thus potatoes, tomatoes, peppers and lima beans became part of the world's diet. Some of these were heat-loving plants which could not be grown with simple Indian tools in the countries that are now the United States and Canada. However, the three great Indian foods—corn, beans and squash—soon spread northeast and west, wherever the climate would permit. Next to the cornfields there were often patches of tobacco, whose incense, the Indians thought, was pleasing to the gods. In the warm southwestern United States, cotton was grown and woven. In each area, Indians developed a way of life suitable to the climate and the materials they could find.

American Indian tribes are as different from one another as are the nations of Europe. Their stature varies from tall and thin to short and plump. An Indian's nose may be hawk-shaped or button-shaped and his skin anywhere from black to pale tan. Some live—or did live—in skin tents, some in stone houses of many stories. They speak over a hundred different languages. Nevertheless, students have organized these languages into six great families, each with sounds and grammar of the same general kind, even if the speakers of one tribe within a language group cannot understand the others. The six language families are Algonkian, Iroquois, Muskhogean, Siouan, Athapascan and Eskimauan.

Indians also are classified in several other ways. One is according to culture areas—that is, the kind of country they inhabited and their way of life there. Thus we speak of the Indians of the Eastern Woodlands. This is the stretch of country from the Atlantic to the Mississippi and up to the barren grounds of north Canada. In Indian days, it was all one great forest, furnishing animals, birds and fish in plenty. The red men padded through it on narrow forest trails or they used canoes, for the whole country was threaded with streams. In the valleys beside these waterways they had their cornfields, tended mostly by the women, for the men were busy hunting and fighting. These woodlanders made everything they

NAVAHO WOMEN are noted for their long, ebony-black hair, which they keep gleaming with "brushes" of stiff grass. The Navahos love color, especially tawny, desert hues.

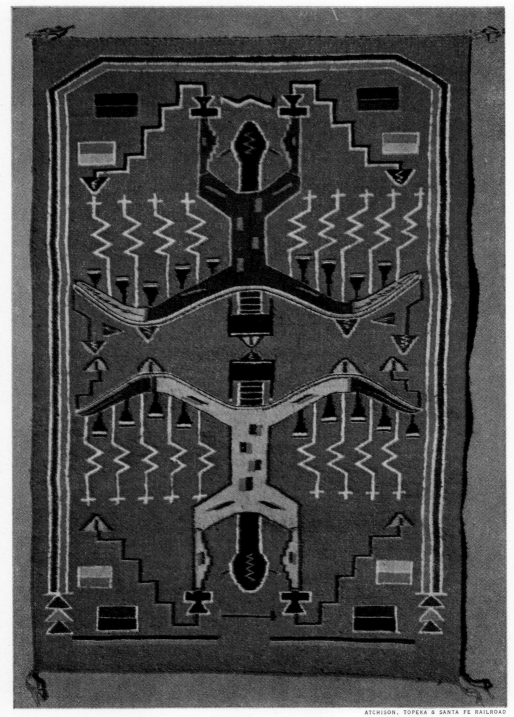

THUNDERBIRDS FLASH LIGHTNING from their wings in a modern Navaho blanket. Early patterns were mostly geometric, with few colors except the natural grays of the wool.

could out of forest products. Canoes were of birch bark in the north, elm bark in New York State, and dugouts were made of cypress in the south. Houses were of poles, covered with bark or matting. Buckets and dishes were of bark or basketry, with a little pottery. Clothing was of buckskin; bedding, of furs.

Here lived people of Algonkian speech. In the north, they were hunters only, like the Naskapi of Labrador. Farther south, they combined hunting and agriculture. It was Indians of the Narragansett tribe who taught the Pilgrim Fathers how to plant corn, four kernels to a hill, with beans between the rows and fish heads for fertilizer. To them or other Algonkians the white man owes the use of the clambake and the bean pot, which were Indian methods of fireless cooking. With their help the newcomers developed such foods as johnnycake, buckwheat cakes, maple syrup and succotash.

"Succotash" is an Algonkian word. So are "woodchuck," "wigwam," "wampum," "squaw," "papoose," "tomahawk," not to mention scores of place names. How bare a map of eastern North America would be without such words as "Massachusetts," "Connecticut," "Adirondack," "Allegheny"!

Along Lake Superior and the upper Mississippi were other Algonkians, pushing their canoes through the wild-rice swamps or tending their cornfields in river valleys. School histories sometimes show pictures of their wigwams, which were frames of poles in domed or gabled shape, covered with bark or matting laid on like clapboards. More often they show pictures of the warriors, with heads shaved except for the scalp lock, and faces painted. Less often they tell about the Algonkian belief in the Great Spirit, the manito, whose power was in every plant, bird and animal. Boys approaching manhood used to go into the forest to fast and pray, hoping for a vision of one of the powerful beings that would give faith and strength for the life of hardship that lay ahead.

Indian Reservations as Homes

Algonkians of the east have now mixed with the general population or they are on small reservations, under the states, in the United States, not the national Government. Algonkians in what is now the Middle West sold their lands and moved still farther west of the Mississippi to reservations.

Another woodland group was the Iroquois, the famous Five Nations—Mohawk, Oneida, Onondaga, Cayuga and Seneca. Later they were joined by their kinsmen the Tuscarora from the south. In fact, the Iroquois seem to have moved up from the south, fighting their way through the Algonkians until they were established in the green valleys of central New York, while the Huron and others speaking the same language built their villages on the Canadian side of Lake Huron.

Even though women did the farm work, the Iroquois cornfields were always im-

SANTA FE RAILWAY

A MEDICINE MAN obeys the Navaho belief that sand painting must be erased by sundown.

AMERICAN INDIANS

mense. Among their cherished deities were the "three sisters"—the spirits of corn, beans and squash—as well as beings of the mountain and forest. The Iroquois did not go out alone to pray to their spirit helpers, as did the Algonkian hunter. The Iroquois held dignified ceremonies within the community long house, when dances were performed and tobacco incense was burned. Some of those dances are performed today, both in New York State and in Oklahoma. Sometimes masks were carved from living trees, to represent the wood spirits who could both cause and cure disease or injury.

The Fortress of the Iroquois

Iroquois villages were stockaded for defense and, inside, were the famous long houses. These were bark-covered dwellings almost as long as a Pullman car and divided into compartments in the same way. Each compartment was the home of a family, whose mother was its head; her husband left his own mother's home to come and live with her. Apparently that arrangement, used in many parts of the world, works just as well as the white man's way, with the father as head of the family. Families were gathered into clans, clans into tribes and the five (later six) tribes into a primitive United Nations. Their great assembly met once or twice a year to "take up the hatchet" or to "bury the hatchet"—that is, to make war or peace.

No wonder the Iroquois were able to conquer the neighboring Algonkians. They built up something like an empire which did not collapse until the American War of the Revolution. At that time their nations could not agree on which side to take and so their unity was broken. The Iroquois now live like whites, some on reservations in the United States and some in Canada. The Caughnawaga group of Canada have become famous as structural-steel workers.

The rich warm land that is now the Southern states was occupied by other Indians, farmers and hunters too, with a long history of settled living. These were the Muskhogeans: Creek, Choctaw,

NATIONAL MUSEUM OF CANADA

THE TSIMSHIAN INDIANS in British Columbia have a family totem outside every home.

Chickasaw and Seminole. One Iroquoian tribe was among them, the Cherokee, who must have been dropped off as their kinsmen were moving northward at an earlier period.

These people are known in the United States Indian Office records as the Five Civilized Tribes, and they merit the name. It was probably the ancestors of some of them who built the famous mounds found in Ohio and for hundreds of miles north and south. Living closer to Mexico than

SAND PAINTING is an art of the Navaho medicine man. It is part of a religious rite performed to heal the sick. Though the artist drops sand freehand, he follows strict rules.

THE ZUNIS, of western New Mexico, are skilled at making silver jewelry, which is often set with turquoise. Though only simple tools are used, designs may be quite elaborate.

THE SONS AND DAUGHTERS of Cree Indian trappers attend the school at Lac La Ronge, Saskatchewan, in northern Canada. In this classroom, the children learn their three R's.

the northerners, the Muskhogeans may have learned to plant earlier. They built large villages, which the whites called towns, with houses of logs covered with earth for winter and spacious arbors for summer. Their women made fine pottery and baskets. Their men were fierce warriors whose clan signs on a blazed tree boldly told a victim just whom he must seek in vengeance.

The Civilized Tribes, like the Iroquois, had mother clans. These were gathered into towns with numbers of officials and a chief whom the whites called "king." On ceremonial occasions the king with his warriors and councillors sat on wooden benches around the town square. When new fire was made at the new year, in summer, they all took an emetic to purify themselves, made peace with each other, forgave debts and generally prepared for a co-operative year. The Five Civilized Tribes have made an admirable record as American citizens. Their land was bought in the 1830's and most of them moved to Oklahoma. There they set up towns with the old names. They ran their own government and schools. Some of them already had constitutions; and Sequoya, the famous Cherokee leader, had provided his tribe with a written language. In 1907, when Oklahoma became a state, the tribal governments were dissolved. Many members by that time spoke English. Cherokee have become successful as lawyers, doctors, teachers, a vice-president of the United States (Charles Curtis) and, perhaps best known of all, the famous part-Cherokee Will Rogers.

Some tribal remnants have not moved. A few Cherokee cling to the Great Smoky Mountains where they have a reservation. Most of the Seminole, long before removal, had run away, as their name implies, to Florida. ("Seminole" comes from a Creek word that means "runaway.") There they fought the United States troops who tried to remove them. They were never really conquered, though they live peacefully today on their reservation in the swampy Everglades of Florida.

Crossing the Mississippi we come to the plains—the Great Plains where once thousands of buffalo roamed like a moving black carpet. We often think of the red men who hunted these buffalo as *the* Indians. Actually, they did not take up their roving life with horse and tepee until after the Spaniards had brought horses to the West, about 1600. Then Indians from many parts of the country pushed out to the rich plains to join the few already established along the rivers. The Cheyenne and Arapaho were Algonkians from the east. Apache came from the far north and Comanche from the west. The Sioux, or Dakota, best known of all, were relatives of the Muskhogeans, from the Mississippi Valley in the central United States.

Buffalo, the Mainstay of Life

Here, on the treeless plains, the main source of supply was the buffalo. Its skins, sewed together, formed the pointed tent which is called by the Siouan word *tepee*. The buffalo's sinews made sewing thread and bowstrings, its bones were tools, its stomach was a bag for meat. Practically all the rest of the animal was eaten, including a great many parts that whites would have refused. These parts were rich in vitamins, however, so that Indians got a fairly well-balanced diet without green vegetables. The meat was sun-dried for winter use into what the whites call "jerky." Pemmican, a Cree word, is dried meat pounded fine and packed in sacks. Add a few berries and greens in season and some buckskin for clothing and the Indian had all he needed. For decoration he wore the famous feather bonnet, though this was a privilege for the few. After the traders came, his clothes were richly embroidered with beads.

Since the plains Indians had no towns like the woodland people, they did not need much government. In the winter, when there were few buffalo, little groups camped by themselves. In summer, a whole tribe might get together for the great hunt and for the sun dance when they prayed for the welfare of the world. Individual warriors, also, went out to implore the Great Mystery for success in war, which was the plainsman's whole career. It was not scalps they tried for but the honor accorded daring acts—to steal an enemy's horse from right outside his tepee, or to touch a dead enemy while his comrades fought around him. These Indians are now farming and running cattle on reservations in the west of the United States and in Canada.

In New Mexico and Arizona there are Indians still living in the style of centuries ago, wearing their old costumes, at least for ceremonies, speaking the old languages and observing the old religious rites. In the south of Arizona are the Pima, some of the first Indians in the United States to practice agriculture. Long before the whites came, they were building irrigation canals as much as sixteen miles long. In the western desert are the Papago who still pick the fruit of the giant cactus and drink its juice as a rite to make their crops grow.

Along the Rio Grande and west into the desert are the Pueblo peoples. *Pueblo* is Spanish for "village" and so the conquistadors named them, three hundred years ago. Their warm, sunny country provides neither forest nor buffalo plains, only sand and rocks with a few scrubby trees and desert bushes. What were the Indians to use for house and utensils? Their answer was the earth itself. Houses were made of stone or else adobe, which is the hard desert earth shaped into home-made bricks. There is summer rain in this country, so that it is possible to raise corn if it is the right variety. The Hopi have tough little plants that they place in the earth twelve inches deep and that give corn as in the other Pueblos, yellow-blue black-red-white, and speckled.

Holiday Costumes of Pueblos

On feast days, you may see Pueblo Indians dressed in cotton that was grown and woven by themselves or in wool from the breed of sheep brought by the Spaniards. In some Pueblos the men wear their hair long, flowing beneath a bunch of parrot plumes. In others they wear strange masks, representing the rain spir-

A GROTESQUE MASK is carved by a twentieth-century Iroquois. His forefathers used such masks in healing ceremonies. They were cut from a living tree and usually painted.

A SEMINOLE ROBE is created on a sewing machine. Once the thousands of tiny pieces were seamed by hand. Some Seminoles still live in their old home, the Florida Everglades.

THE NAVAHO INDIANS make the walls of their houses from poles and the round roofs from earth. Near Lukachuka, Arizona, a family gathers outside a typical Navaho dwelling.

its on whom they call. As you walk among their neat little houses, you buy beautifully decorated pots made by the women or the charming water-color pictures that both boys and girls have begun to paint since they have gone to the government schools and found the new materials.

Out in the desert or the mountains are the Navaho and Apache, speaking Athapascan, a language from the north. Students think that they wandered down from the wild north centuries ago. The Navaho then mixed with Pueblo people, learned to grow corn and to weave; and when the Spaniards came, the Navaho learned about sheep and horses. Now, scattered over the desert, you may see their little domed huts, made of poles covered with earth, and white sheep and horses grazing near. Beside a hut, called a hogan, sits a woman at her loom, weaving one of the brilliant blankets in stripes and zigzags that are now sold all over the United States. The man, if he has not ridden off to the trading post or to a "sing," the Navaho name for a ceremony, may be making silver jewelry, an art learned from the Mexicans.

The Apache did not stop fighting until long after the Navaho, and many were the skirmishes they had with the United States cavalry. Now they have settled down and, as it turns out, their mountain homes furnish some of the best cattle and lumber country in the Southwest. On their five reservations, they run cattle, sheep and a lumber mill and they are talking about a tourist camp. Even so, some of them find time for the charming open-air ceremony in summer when they inaugurate their girls into womanhood.

North of this southwestern country are the arid lands of California and Nevada where the Indians lived on nuts, seeds, rabbits and any other food they could dig, pick or kill. Most of them have changed rather quickly to the white man's clothing and ways. In Idaho were hunters and fishers, also changing rapidly now that the land is farmed and the rivers dammed. On up into Alaska, we still find hunters, dressed in furs now and following the herds of caribou. Many of these speak Athapascan, like the Navaho and Apache.

The coast of Alaska, of western Canada, of Washington, Oregon and northern California is a different place. Here is a narrow strip where the Coast, or the Cascade, Range attracts plentiful rain. There are, or there were, huge forests of spruce and pine, alive with elk, deer, bear, fox and mink. Cold rushing streams go down to the Pacific and in these, every summer, the salmon came to spawn.

Indians living here were as rich as those on the buffalo plains and they had more variety. The soft, straight wood of the evergreens allowed them to hew out planks, even with stone tools. They built houses that looked almost like an unpainted New England barn. They dug out huge canoes, some able to hold sixty men or to plow far into the Pacific after whales. Even their dishes were made of wood but beautifully smoothed, inlaid with shell or carved and painted in animal shape. The Nootka and Tsimshian, on Vancouver Island and the Canadian coast, made the famous totem poles.

These fishing Indians went barefoot, and their clothing was about the size of a bathing suit, made of shredded cedar bark. For the men, it was a breechclout; for the women, a little fringed shirt or two aprons. They had raincapes of matting and, for great occasions, robes of bearskin or the soft brown sea otter.

Potlatches—Gift-giving Feasts

Their ceremonies must have been magnificent to see. With their woodworking craft, they could make great painted masks, sometimes of a whale, which would open and shut its mouth, or of a raven, which flapped its wings. Of course there was a difference up and down the coast, the northern people having the handsomest products. They gave these away at great feasts called Potlatches when a man would beggar himself for grand effect. Then, of course, his neighbors invited him in turn and he got back the same amount or more. It was a primitive form of investment. Indian fishermen now use outboard motors or they stay ashore and work in canneries.

Finally we come to the very rim of the continent. All around the northern seacoast, from southern Alaska to southern Greenland, lived the Eskimos. It must not be thought that they all lived in snow houses. Eskimos were of two kinds. The Water Eskimos lived on the more southerly coast, where they could be out in their skin canoes, or kayaks, for at least half the year. Some of these were in Alaska and some in Greenland. Their houses were of driftwood or of stone, covered with turf. In summer, they lived in skin tents, while they put out in canoes for hunting. One-man canoes, or kayaks, were used for seal, sea lion and birds. The big boat, or umiak, was for whales which they harpooned and towed in to shore. In autumn, they went inland hunting the caribou, which furnished their clothing. In winter, when the waters were frozen, they made trips by sled to hunt or trade.

Ice Eskimos also hunted in summer and lived in tents. In winter, they fished for seals through holes in the ice. This was the time when snow houses were necessary, since they camped on the ice itself, with no stone or tree within miles.

All the Eskimos wore beautifully sewed clothing, made of caribou hide, with boots, hoods and mittens. All had excellent tools, made of walrus ivory and such scraps of wood as they could get. Some in Alaska had pottery, but mostly their utensils were of stone, skin and wood. Their lamps were shallow dishes scooped out of stone and filled with seal oil.

All this is in the past. Long before World War II, white people had begun to live in Eskimo country as traders, teachers and government officials. Eskimos were moving from their old-style houses, changing to modern clothing and being employed in cannery and other work. The war speeded up the change, for men came in planes to the remotest parts of the Arctic for survey and study. In Alaska all Eskimo children can now go to school, even though they have to be flown a hundred miles by plane. The old hunting life is practically at an end, and future Eskimos will live very differently from their ancestors.

In the United States, all Indians are now citizens, eligible to vote, to receive social-security payments and to be drafted for the armed forces. In the war they did excellent service.

In the 1950's the number of Indians living on reservations in the United States was about 400,000. A reservation is a

A BEADED SIOUX COSTUME. When the white man appeared on the Plains, he brought fine beads with him. The Indians quickly substituted them for quills in dress decoration.

TOTEM POLES and totemic figures of the Northwest. The topmost carving, with downturned nose, is a hawk symbol. At the left is a "welcome figure," with outstretched arms.

tract of land set apart by treaty for Indian use in return for other lands given up for settlement by whites. Indians need not occupy this land unless they wish, but the reserving of it by the Government is in payment of a debt. A recent law provides that they may trade or bequeath it to other Indians but not to whites. Thus each group is assured of some permanent place of its own in the country. When they first moved to reservations, the groups were generally provided for in the form of rations—tools, seeds, clothing and the like. Often such supplies continued long after the period mentioned in the treaty. Now most of those periods have run out. However, a court of claims has been set up so that each tribe can make sure it has received all the money promised. Several million dollars have been claimed and paid already to tribes in several states.

Life on the Reservations

On each reservation in the United States the Government provides roads, schools, doctors, hospital service, often farming advisers and social workers. An agent is in charge with an administrative staff on which Indians are employed as fast as they can qualify. Where Indians live near public schools, the Government arranges with county or state for their attendance along with the whites. Promising students wanting education beyond high school can get a government loan. Until the war, Indians were often reluctant to attend school. Now, since many have traveled with the armed forces, they are awake to the possibilities. Over 23,000 were in school in 1953, and the Government was having a hard time to provide facilities enough.

By the Indian Reorganization Act of the United States, any tribe or part of a tribe may incorporate—as a white man's village does—may adopt a constitution and elect officers to look after local affairs. It may plan its own improvements and borrow money from the Government to carry them out. In this way many groups have bought cattle and machinery or sent young men for higher training. Reservations have their own Indian-style courts which try minor offenses, while major crimes go to the federal courts. By a recent law, the different states may, if they wish, take over jurisdiction from the Federal Government. The Indians themselves have organized a congress of all tribes which meets yearly to discuss problems and make suggestions to the Government.

On all reservations there are missionaries of one denomination or another, and many Indians are baptised Christians. Sometimes they maintain the old ceremonies side by side with the new ones. Some still prefer to find their comfort in the ancient dance and song. Indians are at all stages of change from hut dwellers who speak no English to the well-dressed college man who may be an official in the Indian Service.

In Canada, Indians do not vote while they live on the reserves. These are the equivalent of reservations in the United States and are managed by the Department of Citizenship and Immigration, Indian Affairs Branch. If an Indian can prove that he is competent to handle his affairs in the modern manner, he may move off the reserve, be given the right to vote and live like any other citizen. In 1950, the population on the ninety reserves was 136,407. The Canadian Government takes care of roads and general improvements. It builds and maintains the schools but these are staffed by the various churches. Some Indians go to the regular provincial schools. Medical services are given by the Department of National Health.

Needy Indians are given allowances by the Indian Affairs Branch, but the Government makes every effort to help them become self-supporting. Many in the north are hunters and trappers, and projects have been instituted to increase the supply of beaver and muskrat. Farther south, they are farmers and day laborers. Assistance has been given them in agriculture, animal husbandry, fishing, logging and lumbering. As in the United States, an Indian band may elect a council and, through it, borrow money from the government trust fund for group operations.

By RUTH M. UNDERHILL

CANADA, MONARCH OF THE NORTH

The Development of a Colony into a Nation

For all its years of history, Canada has retained the healthy eagerness and vigor of a young and growing nation. Originally settled by peoples of diverse languages, religion, customs and laws, the country has developed a distinctive Canadian consciousness, and its citizens look back upon their past with pride and face the future with confidence. This chapter will give a general background for an understanding of Canada as a whole. The chapters that follow will fill in the details about the various provinces that stretch all the way from the Atlantic coast to the shores of the Pacific Ocean.

CANADA is the nation that occupies the northern half of the North American continent, except for Alaska, which belongs to the United States. Although its land area slightly exceeds that of the United States, much of it is of limited use for human occupation because it lies so far north. Despite this disadvantage, Canada has made the most of its more habitable regions and has found ways of extracting great wealth from those areas that were once looked upon as wasteland.

Canada is no longer the British colony it was for many years. Politically it is a constitutional monarchy like Great Britain or Australia—a completely independent nation. It can make war and peace of its own accord, it has its own ambassadors to other countries, it makes its own treaties and has its own citizenship laws. Although the Queen of Great Britain is also the Queen of Canada, she acts, in Canadian matters, only as advised by her Canadian ministers or by the Parliament of Canada. The British Government no longer has any say in Canada's affairs.

The country's political organization closely resembles that of the United States, with the powers of government divided between the national authority and the local government in the ten provinces. In addition to these provinces it also has two territories in its extreme north, the Yukon Territory and the Northwest Territories, both of which are administered by the national Government. The area of Labrador, which is on the mainland, is part of the province of Newfoundland.

The heart of this enormous half-continent was opened and exploited along two great waterways. The water routes of the St. Lawrence River and the Great Lakes were in the hands of the French until 1759; those of Hudson Bay and the rivers emptying into it were from the early seventeenth century more or less controlled by the British. The French-controlled waterway led to one of the most fertile and productive areas of the continent and gave access to the whole interior plain. The Hudson Bay waterway led to nothing but a vast and inhospitable country which had, however, almost unlimited supplies of furs. The nature of the land under French domination made its rapid settlement possible, and by 1759 France had a well-established and prosperous colony. The Bay country, on the other hand, remained a desolate tract in which the only habitations were the posts of the adventurous fur traders.

When the British captured Quebec and put an end to the French empire in North America, the French colony had some 60,000 inhabitants. To guarantee the loyalty of these people to its rule, the British Government permitted them to maintain the French language and the civil laws of France and to follow the Roman Catholic religion. The descendants of these 60,000 have grown today into a body of nearly 4,000,000 French-speaking Canadians. Most of them live in the province of Quebec, where their language predominates and where the educational system provides for both Roman Catholic and Prot-

PARLIAMENT BUILDINGS—ON A BLUFF ABOVE THE OTTAWA RIVER
The Peace Tower rises impressively over the gables of the Center Block (right) of the Parliament Buildings. The West Block is in the background; the War Memorial, on the left.

estant schools. Schools of each denomination are controlled by the respective religious authorities, and are supported by taxes levied upon their followers. This Separate School system exists also in Ontario and in the three prairie provinces of Manitoba, Saskatchewan and Alberta. However, the special rights of the French language and the French civil laws are confined to Quebec. French is, of course, an official language in the national Parliament and in all federal courts.

This, then, explains the persistence in a single nation of two distinct cultures, two distinct stocks, which intermix only slightly, and two languages. It is a condition deeply imbedded in history, and on no other terms could the nation have been held together.

For some years after the taking of Quebec the English-speaking inhabitants of the new British colony of Canada consisted of a few traders and government officials and the occupying troops. But after the American War of Independence a large number of those who had supported the British in that conflict moved north and settled in the colonies of Nova Scotia, New Brunswick and Ontario. At length the colonies under Britain came to have more English-speaking people than French. By the time of the War of 1812 it was clear that the French population preferred to remain under the British flag rather than join the new republic of the United States.

Half a century later the development of railways had made communication easier, thereby cementing common interests of the colonies. Although still organized in several different units, the colonies began to consider forming a single British dominion which would extend from the Atlantic to the Pacific. In 1867 the British North America Act created a new political unit consisting of the province of Canada (combining Upper and Lower Canada, now Ontario and Quebec), Nova Scotia and New Brunswick, to "form and be one Dominion under the name of Canada." Provision was made for the subsequent inclusion of Prince Edward Island, British Columbia, the territory of the Hudson's Bay Company, and Newfoundland. All of these areas have since been incorporated, although Newfoundland did not come into the confederation until 1949.

The land area of this vast country covers 3,606,551 square miles, of which only 547,946 are considered suitable for agriculture, and 1,345,840 square miles are under forest. The remaining 1,712,765 square miles are made up of muskeg and rock, but while unproductive on their surface, they contain immense mineral wealth.

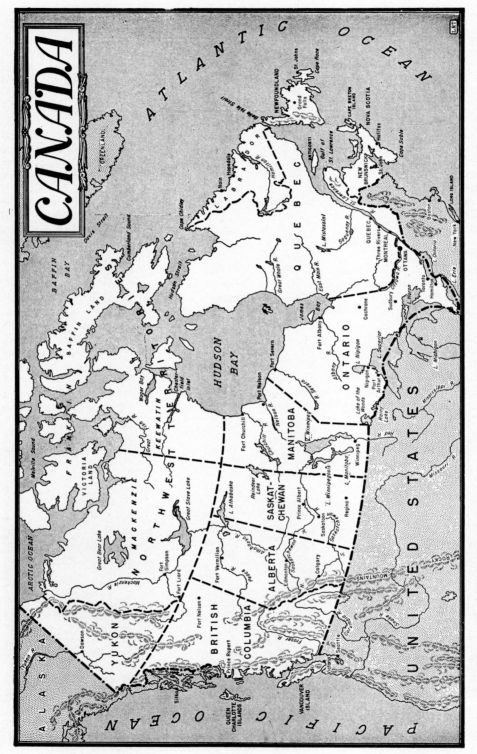

A LAND BETWEEN TWO OCEANS—FROM CAPE RACE TO THE QUEEN CHARLOTTE ISLANDS

Until recent years the exploration of this rock-and-muskeg area was delayed by the high cost of transportation, for the building of railways and the making of roads across its barren face was very difficult. The advent of the airplane and new methods of testing for ore bodies from the ground's surface has altered the picture, however. The mineral stores and possibilities of the shores of the Arctic will soon be as well known as those of long-settled and populous areas.

Most of the rock-and-muskeg land and a considerable part of the forest area lies in a geological formation known as the Canadian Shield or Laurentian Plateau. This is a shield-shaped region of denuded ancient rocks which embraces Hudson Bay, comes down to the shores of Lake Superior and Lake Huron, and occupies half of Canada's total land area. Scoured by glacial action, its rugged and irregular surface has not had time to develop a thick, fresh cover of soil, except in a few places such as the Clay Belt in northern Ontario and Quebec. Its trees are small and, until the rise of the newsprint industry based on pulpwood, its forest land was regarded as a permanent wilderness; but the combination of suitable wood and streams for floating the logs, abundant water power and mineral wealth, has given the Shield great economic importance.

West of the Shield lie the prairies, the Canadian extension of the great, fertile central plain of North America, an expanse of rich soil where wheat growing extends far north into the Peace River country of Alberta. West of the prairies are the Rocky Mountains, the backbone of the continent, and beyond them are the Pacific coast and Vancouver Island, with great valleys of good land between forested mountains and a climate warmed by the Japan Current.

East and south of the Shield is the St. Lawrence Plain, gently sloping agricultural country containing half of Canada's population. Most of the nation's manufacturing has been drawn here by the transportation facilities of the St. Lawrence River and the Great Lakes and also by the cheap electric current produced by the waterfalls of the region. Because of the conformation of the Great Lakes, a part of this plain extends far down into

CANADIAN PACIFIC RAILWAY

THE CITADEL, A RELIC OF BOTH FRENCH AND BRITISH RULE

High above the St. Lawrence in the city of Quebec, zigzag the fortifications of the Citadel. Built in the 1820's, the present walls enclose earlier strongholds of the French.

HER MAJESTY AND PRINCE PHILIP ON A VISIT TO CANADA
Queen Elizabeth II and the Duke of Edinburgh at the picturesque lodge in the Laurentians, in Quebec Province, where they rested for a few days during their whirlwind tour.

CANADIAN GOVERNMENT TRAVEL BUREAU

GREAT DIVIDE GATE ON MOUNT HECTOR IN YOHO NATIONAL PARK
The massive timber gate marks the border between Alberta and British Columbia and part of the Great Divide. On one side all streams flow to the Atlantic; on the other, to the Pacific.

CANADIAN PACIFIC RAILWAY

CLATTERING CHUCK WAGONS RACE AT THE CALGARY STAMPEDE
Raising great clouds of dust in a fond salute to rough and ready frontier days, men of the Prairie Provinces join in a hard chuck-wagon race around the track at Calgary, Alberta.

the latitude of the United States and has a climate mild enough for the growing of tobacco and many delicate fruits.

The provinces by the sea lie still farther east of the Canadian Shield. They consist of two islands, Prince Edward Island and Newfoundland, the peninsula of Nova Scotia and the largely forested mainland area of New Brunswick. Geologically these form an extension of the Appalachian mountain chain of the United States, but their mountains have weathered to the point where they contain (except in Newfoundland) a good proportion of arable land and high-grade forest. The waters along their coasts are excellent fishing grounds. Industrially the Atlantic provinces are at some disadvantage because of their distance from the chief markets of the country.

The growth of Canada was slow until it was shown that wheat could be very successfully grown in the prairies. Large-scale immigration began around 1900. In 1901 the country's population was a little over five million, of whom less than half a million were in the prairies; today the total is about fourteen million, of whom the prairies have over two and a half million.

Canada's annual production of wheat ranges from 300 to 560 million bushels, and the export of wheat and wheat flour is exceeded in money value only by that of newsprint paper. The other important exports are lumber and wood pulp, nickel, copper, asbestos, zinc, aluminum (made from foreign ores but with cheap Canadian water power), agricultural machinery, cattle and fish.

It can be seen that the economy of the country is highly specialized in the production of a few staple articles for which Canada has natural advantages. Since it needs great quantities of subtropical food-

N. Y. NEW HAVEN & HARTFORD RR. CO.

NIAGARA, AT THE BRINK OF THE GIGANTIC HORSESHOE FALLS
In a magnificent, irregular arc above a magic spray of mist, Horseshoe Falls sweeps toward wooded Goat Island. Beyond are Rainbow Bridge and the turgid spillings of American Falls.

NORTHERN SASKATCHEWAN—A URANIUM MINE UNDER CONSTRUCTION
Above Lake Athabaska in the booming town of Beaver Lodge, building supervisors check blueprints. The laced pile of steel rigging is the skeleton of a shaft for a uranium mine.

stuffs—raw cotton, machinery, coal, sugar and rubber—the volume of Canada's foreign trade is exceptionally heavy in relation to the country's population. Some of the articles now imported could never be produced at home. However, in the past few decades Canada has gone a long way in the mass production of machine-made goods for its own use, and in the automobile and farm-implement business it has built up a substantial export trade. It is one of the world's chief sources of radioactive minerals.

The new population that has come into the country since 1900 has been drawn from many different sources. The French were the first people to come in large numbers to Canada, and for many years the Irish ran a close second. However, the Irish are now exceeded by both English and Scots. Germans, Ukrainians and Netherlanders make up other important groups. About one per cent of the population are native Indians and Eskimos. There are even fewer incoming Asiatics.

In some respects Canada's political structure resembles that of Great Britain. In others it is like that of the United States. Since the Queen cannot be in Canada while she is performing her royal duties abroad, she is represented by a governor general, who is selected by the Canadian Government. With the appointment of the Right Honorable Vincent Massey in 1952, this position is filled for the first time by a Canadian citizen. The governor general can act only in accordance with the advice of the Government, which is headed by a prime minister. The Government remains in power only so long as it can retain the support of a majority of the members of the House of Commons, the lower, or elected, house of Parliament. If a Government loses the confidence of the Commons, it must resign or ask that Parliament be dissolved and a new House of Commons elected. Elections are not held at a fixed date, but any Parliament that is not dissolved earlier expires five years from the time of its election, and a new House of Commons must be elected.

To become effective a new law must pass both the Commons and the upper house, or Senate. Members of the Senate are limited in number (there is a fixed quota from each province) and hold their seats for life. The Senate frequently amends the laws passed by the Commons, often with good effect, and the Commons

CANADIAN NATIONAL RAILWAY

A CONCERT IN THE PARK AT VANCOUVER, BRITISH COLUMBIA

In the forested setting of Stanley Park in the heart of Vancouver, hundreds enjoy a concert of light classical music, one of many cultural programs in the bustling western city.

usually accepts the amendments; it is rare for the Senate to vote down entirely a measure upon which the Commons is strongly determined. If that were to happen, the Government would ask for a dissolution of Parliament and be guided by popular opinion. The Government of the day fills Senate vacancies as they occur.

Except for the fact that membership in the Senate is not hereditary as it is in the English House of Lords, these political arrangements closely parallel those of Great Britain. In their internal structure, however, Canada's political parties resemble those of the United States. They hold conventions to select the party leader and to adopt a platform; when a party has a majority in the House of Commons, its leader becomes prime minister. The prime minister then chooses the persons who are to form his Government, in other words, his Cabinet. With the exception of one or two who may be in the Senate, these persons must be members of the Commons or must be elected to that House within a reasonable time.

Party candidates are chosen by a convention of party supporters in each district or province. Party headquarters for the province usually has a good deal to say in the selection, but it does not always dominate the convention.

Government in the provinces is headed by a lieutenant governor who is the equivalent of the governor general in the national Government and who is appointed by the national Government. He has the power to disallow or veto provincial legislation or he can reserve a measure for consideration by the national Government. His power of disallowance is rarely put to use unless the law is believed to be beyond the powers of the province. The question of whether a law is within the powers of the legislative body that enacted it, be it national or provincial, is settled by the Supreme Court of Canada.

By B. K. SANDWELL

CANADA: FACTS AND FIGURES

THE COUNTRY

Occupies the upper half of the North American continent and adjacent islands, excepting Alaska which is owned by the U. S. The country has ten provinces and two territories, a total area of 3,606,551 square miles and a population in excess of 14,010,000. Although its original settlers arrived in the 16th century, the nation has achieved more than 50 per cent of its economic growth in the past 50 years.

GOVERNMENT

Canada is an independent and sovereign state within the British Commonwealth. Executive power is vested in a governor general who is assisted by a prime minister and a cabinet. Legislative power is entrusted to a parliament which consists of a 102-member Senate and a 262-member House of Commons. All British subjects of either sex, who are over 21 years of age and have resided in Canada for a year, are eligible to vote.

COMMERCE AND INDUSTRIES

While the country still depends largely on agriculture, it is rapidly being dominated by industry. The principal crops are wheat, oats, barley, rye, potatoes and a wide variety of fruit. Other prolific sources of income are furs, fish, wool, lumber, dairy products, hydroelectric power and minerals, particularly nickel, coal, asbestos, petroleum, gold, silver, copper and uranium. Today, however, much of Canada's revenue comes from its manufactured products which include newsprint, food, beverages, textiles, chemicals, machinery, transportation equipment and electrical apparatus. Nearly every country in the world trades with Canada, but the United States is its outstanding customer, buying more than 60 per cent of its exports.

COMMUNICATIONS

There are 57,997 miles of railway; 162,000 miles of surfaced road; about 3,000 miles of inland waterways; and 28,000 miles of scheduled airlines. The nation also has some 410,000 miles of telegraph lines; 63,000 miles of submarine cable; and approximately 3,000,000 telephones and 2,400,000 licensed radio sets.

RELIGION AND EDUCATION

Canada has no state church as 49 per cent of its population is Protestant and 43 per cent is Catholic. Educational enrollment comprises about 2,231,000 students in 31,150 provincially-controlled schools; 141,000 in 1,170 private schools; 107,000 in 150 colleges and universities; and 25,000 in 435 full-time and part-time schools conducted for Indians by a Federal agency.

CHIEF TOWNS

Ottawa, capital, has a population of 202,045; Montreal, 1,021,520; Toronto, 675,754; Vancouver, 344,833; Winnipeg, 235,710; Hamilton, 208,321; Quebec, 164,016; Edmonton, 159,631; Calgary, 129,060; Windsor, 120,049; London, 95,343; and Halifax, 85,589.

The Atlantic Provinces
Lands Rich in the Traditions of the Sea

Canada's four provinces on the Atlantic coast are often referred to as the Maritime Provinces because they project into the sea. Their dependence on the sea and its resources has been responsible for their developing a character that is quite different from that of other Canadian provinces. Their history is largely a thrilling saga of hardy fishermen who pitted their skill and wits against the mighty Atlantic. Because the Maritimes have been less affected by industrialism and immigration than other parts of Canada, they still preserve much of their original charm, and many old customs survive.

FACING the North Atlantic Ocean are Canada's four smallest provinces, Nova Scotia, New Brunswick, Prince Edward Island and Newfoundland with its large dependency, Labrador. These are the eastern maritime provinces, whose people live near the sea and to a great extent from the products of the sea. What might be called the outward coasts of Newfoundland and Nova Scotia are washed by the Atlantic, while the inward coasts of the four provinces (excluding Labrador) help form the boundaries of the Gulf of St. Lawrence. The tiniest of the provinces, Prince Edward Island—often called simply "the Island"—is set entirely within the gulf.

Since the Atlantic provinces are nearer to Europe than any other part of North America, it is not surprising that they were among the first regions to be reached and settled by Europeans. Newfoundland, the newest Canadian province, was visited by fishermen from western Europe very early in the sixteenth century. Sir Humphrey Gilbert eventually took possession of the island in 1583 in the name of the English Crown, which made it the first British colony in the New World. Permanent settlement, however, was slow in developing on Newfoundland. The fishermen who landed there to cure their catch did not remain over the winter months.

In 1604 French colonists under Pierre du Guast, Sieur de Monts, made the first settlement on Nova Scotia, which they called Acadie. At first they settled on an island in Passamaquoddy Bay, but they soon moved to the opposite shore where they founded Port Royal, later called Annapolis Royal by the English. Soon afterward the British also claimed this region, and in 1621 it was granted by the British to a Scotsman, Sir William Alexander, who promptly named it Nova Scotia, the Latin for "New Scotland." The area covered the present Nova Scotia and also New Brunswick. This was the beginning of a long struggle between Britain and France for control of the two-pronged peninsula, a struggle that France finally lost. In 1713 Acadie was ceded to Britain. The French remained in Cape Breton where they presently built the fortress of Louisbourg.

This cession made Acadie a British colony, but with a French population wishing to remain neutral in the struggle between the two European powers. Britain and France were each determined to secure the allegiance of these peaceful farmers. The British governor required the Acadians to take an oath of loyalty to the British Crown after war was renewed with France in 1755. When the Acadians refused, it was decided to expel them from the colony, since they were thought to be a serious threat to British authority there. In the late summer of 1755, and during the next eight years as well, numbers of Acadians were put on transports to be dispersed among the English colonies to the south. More than six thousand were thus deprived of their homes. Some lost their lives in shipwrecks, and all were uprooted from their traditional land, although some of them managed to return to Nova Scotia. The sad story of the Acadians has be-

SCHOONERS AT LUNENBURG
Trim vessels of the Grand Banks fleet ride at anchor in a bay of southwest Nova Scotia.

come famous through Longfellow's poem *Evangeline*.

In the same decade in which the Acadians were expelled, their places were taken by an inflow of settlers from New England, which gave Nova Scotia its first large English-speaking population. In those years immigrants from Germany also arrived, and their descendants are still to be found in the county of Lunenburg. About 1775 a large number of Highland Scots came to Cape Breton Island and to the region around Pictou, where many of their descendants speak Gaelic to this day. But the greatest increase in population at any one time in Maritime history came with the American Revolution, when some thirty thousand Loyalists left the rebellious colonies to find new homes in the colonies remaining within the British Empire. So large was the influx that Nova Scotia was divided in 1784 and the mainland north of the Bay of Fundy was formed into a new province called New Brunswick.

During the first half of the nineteenth century the Atlantic provinces grew and prospered. Those were the days of wooden sailing ships, and Maritimers became famous for the craft they built and sailed to all parts of the world. Products of the provinces, especially timber, enjoyed a favored position in the British market, and at the same time trade ties with nearby New England remained close. But after 1850, with the decline of the sailing ship and the coming of free trade in Britain (which meant a loss of preference in that market), the Atlantic provinces entered upon less happy days. Some of their people favored joining the province of Canada to form a new nation, the Dominion of Canada. Such a union would give the seacoast provinces railway connections with the interior. With some misgivings Nova Scotia and New Brunswick entered Confederation in 1867, but Prince Edward Island stayed out until 1873 and Newfoundland did not enter until 1949. Citizens of the Atlantic provinces have sometimes felt that they did not benefit from Confederation as much as did the inland provinces.

The Great Movement West

Certainly the interior of the continent grew much more rapidly than the provinces by the sea. When navigation was open on the St. Lawrence, ships from Europe went directly to the great port of Montreal instead of stopping at Halifax or Saint John. Newcomers from Europe made for the west rather than for the Atlantic provinces, and the sons and daughters of the seaboard areas were also attracted inland. This movement to the interior had a tendency to retard the growth of population in the eastern provinces.

The depression of the 1930's hit the Atlantic provinces a heavy blow. The prices of their export products fell tragically. After 1940, however, better times returned as the provinces entered upon a more diversified economic life.

Now let us look more closely at each of the provinces in turn, starting with Canada's newest province, Newfoundland and its dependency, Labrador. In shape Newfoundland resembles an equilateral triangle whose sides are cut with a great many jagged bays and inlets; thus, there is a plentiful supply of good small harbors. The western shore of the island is marked by a range of high hills from which the land gradually slopes down to sea level on the eastern and southern coasts. The Labrador Current, which flows down the east coast, makes the climate cool, and the growing season is short. The days tend to be gray, with too little sunshine for good farming. A large part of the island is made up of unproductive barren lands and bogs, where neither trees nor crops will grow. Only a fraction of the soil is fertile enough to support agriculture.

The Seafaring Newfoundlanders

The inhabitants are mainly of English and Irish origin. Nearly all of them live on the coasts. Only minor settlements are to be found inland. Travel between one part of the island and another is generally done by sea, although there is a narrow-gauge railway running from Port aux Basques on the southwest coast to the capital city of St. John's on the southeast coast.

Labrador is a wild and rocky land lying east of Quebec and north of Newfoundland. It is uninhabited except along its coasts, where a few thousand Indians and Eskimos live by fishing and sealing.

Fishing dominates the economic life of Newfoundland today, as it has always done. For many hundreds of years men have gone out to the Grand Banks southeast of the island and to other fishing banks in search of cod, herring and other creatures of the deep. Less romantic but more efficient motor-driven trawlers have now largely replaced the sailing ships of former days. Salt cod has long been the leading product of the island's fisheries, but in recent years quick-freezing plants and refrigerator ships have been used to supply the growing market in fresh frozen fish. In one form or another half the population depends upon the fishing industry.

The next most important source of livelihood of the people of Newfoundland comes from the forests of small coniferous trees which cover nearly half the island. Several large pulp and paper companies carry on operations which employ thousands of men. Many hundreds of small sawmills are spread around the island. The lumber they cut is mostly for local use. The value of forest products is higher than that of the fisheries, but the forest industries employ fewer people.

Mining is the third most important industry of Newfoundland. Iron ore is taken from the Wabana mine on Bell Island, just northwest of St. John's, and is shipped to steel mills at Sydney, Nova Scotia. In the center of the island there is also a mine producing copper, lead and zinc. Work is now going on to exploit the extensive iron-ore deposits of northeastern Quebec, which reach into western Labrador.

The eastern location of Newfoundland and Labrador has made them vital links in communications and defense. During World War II, an important airfield was developed at Gander Bay, Newfoundland, and from it thousands of airplanes were ferried to Europe. Since Gander is one-third of the distance between New York and Ireland, it has become a major stopping-off point on transatlantic flights. Another great airport, built at Goose Bay in central Labrador, is a vital link in the air defenses of Canada and the United States.

The Tenth Province Is Born

From 1855 to 1934 Newfoundland was a self-governing dominion. During the depression of the 1930's, however, its economic position became so poor that representative government was suspended in favor of government by a British royal commission. During World War II, the island's position improved greatly and it became possible to restore representative government. This involved a choice of returning to the status of a separate dominion or entering into Canadian Confederation as the tenth province. By a narrow margin, Newfoundland made the second choice.

Since joining Confederation, the provincial government has encouraged European industrialists to establish plants on the island, and the large amount of new capital invested there should bring increased employment and prosperity to the inhabitants. Moreover, Confederation has brought such social benefits as family allowances, which mean a great deal to those with low incomes.

South and west of Newfoundland, across the Gulf of St. Lawrence, are the three traditional Maritime Provinces—Nova Scotia, New Brunswick and Prince Edward Island. They are joined, rather than separated, by the sea. They have excellent harbors from which ships and ferry boats regularly carry passengers from one province to another.

New Brunswick, east of the state of Maine, is less clearly a maritime region than are the other Atlantic provinces; nevertheless it has extensive coasts on the Bay of Fundy and the Gulf of St. Lawrence. Its north coast is formed by the deep indentation of Chaleur Bay, which separates the province from the Gaspé Peninsula of Quebec. New Brunswick is a rugged, hilly land, rectangular in shape. It is largely covered with forest, especially in the northern parts. Through its hills cut the largest rivers to be found in these provinces; one of them, the St. John, which rises in the state of Maine, is the mightiest on the Atlantic seaboard south of the St. Lawrence. This river flows into the Bay of Fundy where it forms an excellent harbor at the site of Saint John, New Brunswick's largest city. Other important rivers are the Restigouche, famous for its salmon, and the Miramichi, both of which flow eastward.

The population of New Brunswick is found mainly along the coasts and in the river valleys. The rugged north-central portion is almost uninhabited. The northern half of the province is predominantly French-speaking and many pioneer settlements have recently been established here. The population of the southern half is mainly British in origin.

Nova Scotia, Peninsular Province

Nova Scotia is connected with New Brunswick by the narrow Isthmus of Chignecto; otherwise it is an island, or rather two islands, because Cape Breton Island is separated from the rest of Nova Scotia by the narrow Strait of Canso.

Like New Brunswick, Nova Scotia is also hilly and rugged, with relatively little land that is good for farming. Hills and uplands cut across the province in a northeast and southwest direction. Only rarely are fertile valleys to be found. The largest and best of these is the Annapolis Valley near the Bay of Fundy shore. The southern Atlantic coast of the province is much rockier than the Fundy and gulf shores and is deeply cut with many inlets. These inlets make fine harbors, the largest of which is at Halifax, one of the world's great ports. Yarmouth, at the western end of the province, is the traditional port for trade with the New England states. The southwestern part of the province, under the moderating influence of the ocean, has a milder climate than any other parts of the Atlantic provinces. Much of the interior is barren and supports little population. This is especially true of Cape Breton Island, whose rugged inland is almost uninhabited. This island is cut almost in two by the Bras d'Or Lakes, which are really mighty salt-water inlets.

Prince Edward Island

Across Northumberland Strait from New Brunswick and Nova Scotia lies the crescent-shaped province of Prince Edward Island. At one place its shores are less than ten miles from the New Brunswick coast. There are no striking physical features on the island, which consists of low, rolling hills well suited for agriculture. It is the most uniformly and completely settled portion of the Maritimes and has sometimes been called "the million-acre farm." The one urban center of any importance is the capital, Charlottetown, situated on a well-protected harbor on the south shore. Like those of all the other Atlantic provinces, the coasts of Prince Edward Island are very irregular and provide many good harbors.

Fishing was the earliest economic activity in the maritime region, and it still is among the most important. In former days the favorite craft of Nova Scotia fishermen was a two-masted schooner with auxiliary oil-burning engines. Such a schooner carried several small boats

THE ATLANTIC PROVINCES

called dories. At the fishing banks the dories, each manned by two men, would leave the schooner and return later to the mother ship with their catch. The most famous of all these schooners was the Bluenose, which for many years was unbeaten in races with ships of its type.

Inshore fishing is usually carried on by motorboats operating within a few miles of the home dock. Its importance has greatly increased in recent years. The Bay of Fundy and the Gulf of St. Lawrence, especially around Prince Edward Island, have been the leading centers of this kind of fishing. The most valuable catch is the lobster, which is caught in wooden traps set in shallow waters near the shore. Lobster has now replaced cod as the leading product of the maritime fishery, for there is a growing market for it in the large cities of the United States and Canada. Oysters and the little herring known as sardines are also caught by inshore fishing methods.

As we have already seen, the Maritimes are not highly favored for carrying on agriculture, except in some of the river valleys and on Prince Edward Island. The latter grows potatoes and raises hogs, which bring a valuable cash return to its farmers. New Brunswick and Nova Scotia also produce crops for export to other parts of Canada and the world, but in general they import more agricultural products than they sell. New Brunswick's leading farming region is in the St. John Valley, where potatoes are grown on a large scale. The Annapolis Valley in Nova Scotia grows a large part of Canada's commercial apple crop. Dairy farming is important in all the Maritimes except Newfoundland. In general, however, the outlook for agriculture is not as bright as in other parts of Canada.

PAN AMERICAN WORLD AIRWAYS

GANDER, AIR GATEWAY TO CANADA AND THE UNITED STATES

Gander Airport, a fog-free spot in eastern Newfoundland, is one of the busiest airfields in the world. Scores of transatlantic liners, like this Constellation, stop at Gander daily.

The land of New Brunswick and Nova Scotia is much more important for its forests than for its agriculture. In fact, almost as many men are engaged in the forest industries of these two provinces as are employed in their fishing industries. The best-forested areas are found in central and northern New Brunswick, although the trees of Cape Breton Island and western Nova Scotia are also important. For many years fishing and lumbering were closely connected, since a main use of the lumber was in building ships for the fisheries. At the present time the cutting of trees for pulpwood is the most important phase of the forest industries.

The Mines of Cape Breton

Nova Scotia is easily the leading province in mining. Nearly half the coal mined in Canada comes from mines in Cape Breton Island and in the isthmus and north shore region. The Cape Breton mines, which are the most important, are at the ocean shore; indeed, some of them extend outward under the ocean floor for as much as three or four miles and are both dangerous and expensive to work. No other mining in the Maritimes approaches coal in importance, although most of the gypsum produced in Canada comes from Nova Scotia.

These provinces are not an important manufacturing region when we look at Canada as a whole. Nevertheless their production of finished goods has increased considerably in recent years. The coal mines around Sydney, in Cape Breton Island, are responsible for the most important single industry—iron and steel manufacture. Iron ore from Newfoundland is processed in the large mills near the coal mines. Industry has also been built on other local raw materials. Pulp and paper are made in New Brunswick, while many fish canneries and processing plants are to be found throughout the provinces. The two leading ports, Halifax and Saint John, are sites for a wide variety of factories, many of which rely upon imported raw materials, such as petroleum and sugar.

Another feature of the economic picture in the Maritimes is the rapidly growing tourist industry. The relatively cool summer climate of these provinces attracts many people from New England, New York and central Canada. Summer visitors enjoy the yachting, salt-water bathing and fishing as well as the quiet, restful atmosphere. Moreover, the many places of historic interest, such as Annapolis Royal, draw numerous visitors year after year.

Finally, we must note the important place that foreign trade has always played in the life of the Maritimes. Since they have the only ports on Canada's eastern shores which are open all year round, these provinces have built up an extensive trade with many parts of the world, especially Great Britain and the British West Indies. Halifax, a vital naval station, expanded greatly during World War II.

Old Centers of Learning

The visitor to the Maritime Provinces will find an intellectual and cultural tradition of long standing. The first college founded in English-speaking Canada was King's College, which was established at Windsor, Nova Scotia, in 1788. It was later moved to Halifax, where it now shares a common campus with Dalhousie University, the largest institution of higher learning in the Maritimes. Another King's College was established at Fredericton, the capital of New Brunswick, in 1800; this college later became the University of New Brunswick. There are also many other small colleges and universities, both religious and non-denominational, in the provinces. An outstanding example of their contribution to the life of the region is the extension program carried on at St. Francis Xavier University, at Antigonish, Nova Scotia.

Over the years the Maritimes have contributed much to Canada's cultural life. In the 1830's Judge T. C. Haliburton of Nova Scotia began to write his famous sketches of "Sam Slick, the Clockmaker," a landmark in the history of North American humor. Two of Canada's finest poets, Charles G. D. Roberts and Bliss Carman, began their careers in New Brunswick.

BY G. M. CRAIG

Facts and Figures are given on page 64.

STATUE OF LONGFELLOW'S EVANGELINE IN NOVA SCOTIA PARK

Evangeline Park in Grand Pré immortalizes the heroine of the well-loved poem. A statue of Evangeline dominates the park, and a replica of the Acadian church appears in the background.

A SWIRLING STREAM in the Kedgemacooge district of Nova Scotia offers an inspiring challenge to angling enthusiasts. Long before the modern fisherman came, the Micmac Indians caught trout in this same stream. At Kedgemacooge Lake nearby is a centuries-old "picture gallery" where Indians, working with a beaver's tooth, cut pictographs on rocks of slate.

COURTESY CANADIAN NATIONAL RAILWAYS

THE RESTIGOUCHE, or Ristigouche, as it is often spelled, is an interesting river, which for many miles of its course divides the Provinces of Quebec and New Brunswick. Some claim that it is the best river for salmon in the world. During the spring and summer it is visited by **thousands** of anglers seeking trout or salmon. Where the river widens and deepens, forming pools, the salmon lurk, and when one takes the fly the angler has need of all his strength and skill. The branches of the Restigouche are also visited by zealous fishermen.

NATIONAL FILM BOARD

APPLE-PICKING TIME IN ONE OF NOVA SCOTIA'S MANY ORCHARDS

The Annapolis Valley of Nova Scotia is world-famous for its apple orchards. Farming is a major industry of the country, and demonstration farms are maintained in many rural areas.

FREDERIC LEWIS

A FAMILIAR STREET SCENE IN NOVA SCOTIA'S FISHING VILLAGES

Most Nova Scotian villagers who are not engaged in farming earn their living as fishermen. Here stacks of lobster pots, buoys and other fishing equipment line the road along the water's edge.

SHIPBUILDING NEAR SAINT JOHN, A MAJOR CANADIAN SEAPORT
Saint John, New Brunswick, at the mouth of the St. John River, is ice-free the year round. It is a shipbuilding and repair center, with one of the world's largest drydocks.

COURTESY CANADIAN NATIONAL RAILWAYS

BLACK HARBOR, off the Bay of Fundy, in southern New Brunswick, is the center of the sardine-packing industry in eastern Canada. In most years this industry is worth more to the province than any other fishery except that of the lobster. The fish sold as sardines on this side of the Atlantic are not the same as the true European sardine but are usually immature herring. These silvery fish swim about in schools, and large numbers of them are caught in a single haul of the nets. Fishing parties, like this one, use seines or trap nets for the capture.

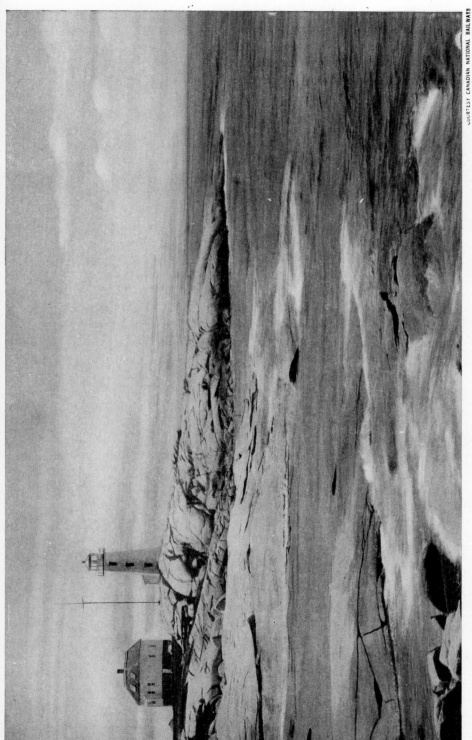

PEGGY'S POINT and Peggy's Cove on the south shore of Nova Scotia, not far from Halifax, are visited every summer by dozens of artists who strive to put upon canvas the ever changing aspects of sky and water and rocks, along these granite shores. In summer this is one of the loveliest spots on the whole coast, but when the winter storms come the exposed lighthouse is a desolate place. Back from the shore there is a varied country of lakes, forests and rivers with a few farmhouses and occasionally a small village, but there are no considerable towns.

TODAY'S LOGS ARE TOMORROW'S NEWSPAPERS

One of Newfoundland's major industries is the manufacture of newsprint. Corner Bay Village, above, is one of the centers of the industry, where pulp logs are converted into paper. Newfoundland has about 25,000 square miles of forest lands that feed the numerous pulp and paper mills. Shipbuilding is also quite important, and more than a thousand sawmills are in operation.

A PULP AND PAPER MILL, CORNER BROOK, NEWFOUNDLAND

Second only to its fishing industry in importance is Newfoundland's production of paper. A mountainous island, Newfoundland is covered with rich forests that supply the mills with a plentiful amount of timber. The paper mill at Corner Brook is the larger of two on the island, and it can produce as much as a thousand tons of newsprint a day.

NEWFOUNDLAND FISHERMEN LOADING DRIED COD NEAR ST. JOHN'S

Fishing is the oldest and most important industry in Newfoundland because the island lies only three hundred miles northwest of the Grand Banks, where the world's finest fishing is found. Fleets of fishing vessels come and go at the excellent natural harbor of St. John's, Newfoundland's capital and its center of commerce and manufacturing.

COURTESY CANADIAN PACIFIC RAILWAY

STRONG OXEN furnish much of the traction in western Nova Scotia, and such a sight as this is common. The meadows are an important source of wealth in the Atlantic Provinces, and the meadows off the Basin of Minas are particularly interesting. Though they are diked and much of the water is kept out by gates, nevertheless, at high tide, they are crossed by an intricate system of channels which are filled with water from the Basin. Hay is, next to wheat, the most valuable crop of Canada and in the Atlantic Provinces is worth much more than the wheat crop.

PHOTOGRAPH BY MACASKILL

THE GASPEREAU VALLEY is a delightful region of apple orchards and fertile farms. For miles and miles carefully tended orchards and farms extend through the valleys of the Gaspereau, the Cornwallis, the Pereau and the Annapolis in the western half of Nova Scotia, where one may "ride for fifty miles under apple blossoms". The scene in spring time when the trees are clothed in pink is too lovely to describe. This is one of the best fruit regions in North America, and the well cultivated farms are hardly less attractive than the orderly ranks of the orchard trees.

MARSHALL STUDIO

MOUNT MORIAH, A FARM SETTLEMENT ON THE BAY OF ISLANDS

Picturesque Mount Moriah is on the hill-rimmed Bay of Islands on Newfoundland's west coast. It is on the Humber River, one of the deep arms of the bay that contains many high islands.

CANADIAN NATIONAL RAILWAYS

AN UP-TO-DATE SILVER FOX FARM ON PRINCE EDWARD ISLAND

Careful crossbreeding on fox farms, or ranches, results in pelts that are wanted in the fashion marts. In the fall, the fur-growing period, each fox is penned alone to protect its coat.

PHILIP GENDREAU

TURNAVIK, LONELY FISHING VILLAGE ON THE COAST OF LABRADOR

The look of the raw Labrador coast seems enough to pierce a man to the bone. Remote Turnavik lives on the profits from the catches of cod that it sends to southern towns for canning.

NATIONAL FILM BOARD

THROUGH A ROW OF BIRCHES IN THE SMALLEST OF PROVINCES

New Glasgow is a modest file of clean wooden homes in the rolling countryside near the Gulf of St. Lawrence coast. Prince Edward Island National Park is but a few miles away.

THE MARGAREE RIVER on the western side of Cape Breton Island, Nova Scotia, empties into the Gulf of St. Lawrence. This river flows through a country widely known for its pleasant scenery, and practically every pool is rimmed with meadowland. Although salmon are found in other rivers and streams in Nova Scotia, they are generally not as large as those found in the Margaree. For this reason the valley is extremely popular with fishing sportsmen. It is equipped with many comfortable inns, all of which cater especially to the angler's needs.

COURTESY CANADIAN NATIONAL RAILWAYS

ALONG THE COAST of the smallest Province, Prince Edward Island. Though no point on Prince Edward Island is more than 390 feet above sea level, the sandstone cliffs along the shores rise rather abruptly in some places. The north shore is almost a continuous beach, and everywhere along the coast the water is safe. Geologically the formation is the same as that of the interior lowland of New Brunswick and the northeast part of Nova Scotia. The Island is sometimes called "the million-acre farm" because so much of the area is under cultivation.

COURTESY CANADIAN NATIONAL RAILWAYS

THE ATLANTIC PROVINCES

ATLANTIC PROVINCES: FACTS AND FIGURES

Nova Scotia, New Brunswick, Prince Edward Island and Newfoundland and Labrador, their shores washed by the Atlantic Ocean, are known as the Atlantic Provinces. Total area, 203,971 sq. mi. Population, 1,618,100.

Nova Scotia is composed of the peninsula proper and Cape Breton Island. Total area, 21,068 sq. mi.; population, 642,600.

GOVERNMENT

Government consists of a lieutenant governor appointed by the Federal Government, and a ministry responsible to a House of Assembly of 37 members. Representation in the Canadian Senate, 10; in the House of Commons, 13.

COMMERCE AND INDUSTRIES

Farming, coal mining, fishing and steel manufacturing are the principal industries. Farm products: poultry, dairy goods and fruit. Minerals: coal, gypsum, sand and gravel, barite, salt, stone and silica. Forests include spruce, fir, hemlock and pine. Cod and lobster fisheries.

COMMUNICATIONS

Railway mileage, 1,420; highway mileage, 15,117; several excellent air fields.

RELIGION AND EDUCATION

The population is three-fourths Protestant. Education is free, compulsory and nondenominational in primary and secondary schools. There are 9 universities and colleges.

CHIEF TOWNS

Population of chief cities: Halifax (capital), 100,000; Sydney, 31,000; Glace Bay, 27,000; Dartmouth, 16,000.

New Brunswick has a total area of 27,985 sq. mi.; population, 515,700.

GOVERNMENT

Government is vested in a lieutenant governor and a Legislative Assembly of 52 members elected for terms of five years. Representation in Canadian Senate, 10; in House of Commons, 10.

COMMERCE AND INDUSTRIES

Wood and paper processing, food and general manufacturing are the leading industries.

COMMUNICATIONS

Railways, 1,836 mi.; highways, 13,178 mi.; 142,000 mi. of telephone wire; 65,613 telephones.

RELIGION AND EDUCATION

Population is about 50% Protestant. Public education is free and nonsectarian; there are 5 universities and colleges.

CHIEF TOWNS

Population of chief towns: Fredericton (capital), 16,018; Saint John, 50,779; Moncton, 27,334; Edmundston, 10,753; Campbellton, 7,754.

Prince Edward Island, the smallest province in all Canada, lies at the mouth of the Gulf of St. Lawrence. It is separated from the mainland of New Brunswick and Nova Scotia by the Northumberland Strait. Area, 2,184 sq. mi.; population, 98,429.

GOVERNMENT

The province is administered by a lieutenant governor appointed by the Federal Government and a ministry responsible to a Legislative Assembly of 30 members. Representation in Canadian Senate, 4; in House of Commons, 4.

COMMERCE AND INDUSTRIES

Fishing and agriculture are important occupations; silver-fox farming extensively carried on.

COMMUNICATIONS

Railway mileage, 286, with ferry connections to the mainland. Telephone-wire mileage, 12,928.

RELIGION AND EDUCATION

The Protestant population is about 55%. Over 20,000 pupils in the more than 460 public and private, primary and secondary schools. There are two colleges in Charlottetown.

CHIEF TOWNS

The population of chief towns: Charlottetown (capital), 15,887; Summerside, 6,547.

Newfoundland is composed of a large island at the mouth of the St. Lawrence River and the eastern portion of the Labrador Peninsula. Total area, 154,734 sq. mi. (area of the island only, 42,734 sq. mi.), total population, 361,416 (population of the island only, 353,526).

GOVERNMENT

Government is under a lieutenant governor, a Cabinet and a General Assembly of 28 elected members. Represented in Canadian Senate by 6 and in the House of Commons by 7 members.

COMMERCE AND INDUSTRIES

Cod, salmon, herring, lobster, haddock, seal and whale fisheries. Crops include hay, potatoes, turnips, cabbage and truck-garden products; livestock numbers about 150,000 head. Chief minerals are zinc, iron ore, lead, copper, fluorspar and limestone. Forest reserves, principally fir and spruce, are considerable; lumber, pulp and paper are the chief items of manufacture. High-grade iron ore has been discovered at the headwaters of the Hamilton River.

COMMUNICATIONS

Newfoundland has 704 mi. of railways with connections to coastal steamer routes. Motor roads, 5,800 mi. Gander and other airports serve transatlantic airlines. 22,000 telephones.

RELIGION AND EDUCATION

Predominant religions are Roman Catholic, Anglican and United Church of Canada. Education is largely in church schools, aided by provincial government; 78,613 pupils. Newfoundland Memorial University is in St. John's.

CHIEF TOWNS

Pop. of chief cities: St. John's (capital), 52,873; Corner Brook West, 6,831; Wabana, 6,460.

QUEBEC, ONCE NEW FRANCE

The Largest Province of Canada

Quebec, known as New France in the pioneering days of North America, was founded by the French about 350 years ago. Though transferred to Great Britain in the latter half of the eighteenth century, it has remained French in spirit and largely in language. As territory has been added to Quebec, it has become the largest province in Canada, and promises to become one of the richest mining centers in the world. Quebec City and Montreal are the two oldest cities of Canada and are among its most beautiful. Though far from the ocean, both are important ports for they are situated on the broad St. Lawrence.

QUEBEC is unique among the provinces of Canada because of its French origins. Eighty-five per cent of all the people who live in Quebec are French-speaking Canadians. Numerous small villages along the banks of the St. Lawrence, with their low farmhouses and high church spires, recall the pleasant villages of France. Many of the farmers are descendants of the original settlers of New France, and they are proud of their French heritage and customs.

When Canada was united on July 1, 1867, the old province of Canada was divided into the provinces of Quebec and Ontario. These two, together with New Brunswick and Nova Scotia, formed the new Confederation. French Canadians, who had spoken their own language and lived under their own laws for a century and a half before the new nation was born, were guaranteed the right to enjoy their traditional civil law and religious privileges in Quebec, as they do to this day.

Quebec is the largest of the Canadian provinces, covering almost a sixth of the total area of Canada. Most thickly settled of the four major regions of the province is the St. Lawrence Valley. The best farming sections and the most important cities are located on the banks of this great waterway. Above the city of Quebec the river averages less than 2 miles in width, but below Quebec it soon widens to 10 miles, and later to 20 miles. Finally, when the great river reaches the Gulf of St. Lawrence, it is 50 miles across between the Laurentians on the north bank and the Gaspé Peninsula on the south.

North of the St. Lawrence rises the Laurentian plateau. The mountains here are among the oldest in the world. Some of the peaks rise to 5,000 or 6,000 feet, but the general elevation is only 500 to 2,000 feet. This glacier-scoured plateau is part of the Canadian Shield, and is dented with numerous hollows and dotted with lakes and streams. Once mere canoe routes, many of the rivers are now sources of mighty electrical energy.

This sparsely populated region is rich in forests and minerals. New towns have sprung up because of mining or lumbering projects. Fish and wild game abound. The Indians in the wild northern parts live by hunting moose, caribou and fur-bearing animals. Along the bleak shores of James Bay, Hudson Bay, Hudson Strait and Ungava Bay a few Eskimos live mainly by seal fishing. Here and there are outposts of the Royal Canadian Mounted Police, occasional radio stations and lone missions.

The northern end of the Appalachian Mountain range comes down to the sea at the tip of the Gaspé Peninsula. One solitary peak, Percé Rock, is the picturesque remains of a sea-girt island cut from the mainland. Along the rugged peninsula nestle the hamlets of fisher folk, dependent for their livelihood on the traditional catches of herring, cod, salmon, mackerel and lobster.

The climate of Quebec along the thickly settled St. Lawrence Valley is pleasant and healthful. Summers are short, but the temperature sometimes rises to 90°F. The long winters are ideal for such winter

COPYRIGHT DETROIT PHOTOGRAPHIC CO.

CHATEAU DE RAMEZAY has had a long and interesting history. Built about 1705 by Claude de Ramezay, governor of Montreal, it was his residence for about twenty years, was then the headquarters of the fur trade and the residence of British governors. Still later it served in turn as the headquarters of the Special Council, as governmental offices, a court house, a law school, a normal school and a medical school. It is now an historical museum. It contains fine exhibits having to do with Indians and with the early days of New France.

ST. LOUIS GATE occupies the site of an older gate in the walls which made Quebec almost impregnable to direct attack. Even the present walls are a reconstruction, as very little of the original work remains. Beyond the walls St. Louis Street becomes Grand Allée on which the dignified Parliament Buildings front. One may ascend the steps to the right of the gate and make the circuit of the walls, about three miles. The unusual sort of cab in the foreground is called a calèche, and was formerly the prevailing type of public conveyance, at least in the summer time.

PERCE ROCK
The sea-girt peak that stands guard off the coast of the picturesque Gaspé Peninsula.

sports as skating, skiing and tobogganing. The air is dry and bracing. In the far north, the climate is subarctic and the thermometer may fall to 30° or more below zero.

The French Canadian farmer is largely self-sufficient. His land is often a long, narrow strip, not more than 500 feet wide, running down to a bit of river front. The typical habitant raises most of the food for his large family, for his hogs supply meat, and his cows milk, butter and cheese. He even raises his own tobacco. He cuts fuel from his own wood lot, and often makes his own maple sugar and syrup. His wife frequently spins and weaves the wool for the family clothing. Practically the only supplies that some families have to buy are tea, cotton cloth, hardware and tools. Many a family lives today in the same stone or heavy wooden frame house built by an ancestor before the yielding of New France to Britain in 1763.

On some small farms the plows are still drawn by oxen, and occasionally one sees a big dog harnessed to a cart laden with cans of milk and bound for the village market. Many large farms, however, have modern farm machinery and follow modern methods of farming. There are many automobiles in rural Quebec, and its country roads have been greatly improved. Along the motor routes the tourist sees hundreds of hooked rugs, socks and other handicrafts offered for sale.

Quebec has English-speaking farmers too, mostly in the western part of the province and in the fertile regions south of the St. Lawrence River. This section, known as the Eastern Townships, along the international boundary, was settled by Loyalists who emigrated from the British colonies south of the border following the American Revolution.

Most English-speaking citizens of Quebec, however, live in the cities and are engaged in business or a profession. Yet Quebec cities, like the countryside, are predominantly French. More than nine out of ten people in the city of Quebec and seven out of ten in Montreal speak French. Next to Paris, Montreal is the largest city in the world where French is the principal tongue. Signs in both languages are everywhere—at the railroad crossings, along the highways, over the stores, at the street corners and in the hotels. Telephone operators, street-car conductors and policemen must be able to speak both languages.

As there are two languages, so there are two religions. As French predominates, so does Roman Catholicism. The doors of the many churches are always open, and a steady stream of worshipers pours in and out from early morning Mass throughout the day. Padres wearing scoop-shaped hats and long, black cassocks, and nuns in their flowing robes are familiar figures on the city streets. Files of children from the schools and convents often walk in pairs along the streets, accompanied by nuns. Many convents are located throughout the province, the largest being in Quebec and Montreal. The convent girls are all dressed alike, in long, black, pleated skirts, black cotton stockings and black hats.

Since both French and English in Quebec wished from the beginning to preserve their identity, they have worked out two separate systems of education. The Roman Catholics have one set of schools, the Protestants another. Each has its own course of instruction, which is supervised by the Roman Catholic or the Protestant committee. Because there is no minister of education, the schools are kept out of politics. A superintendent in charge of the Department of Education is assisted by two deputy ministers, one of whom is the director of Protestant education. From top to bottom the two school systems are distinct and different. Almost all non-Catholics attend Protestant schools.

In the rural parts of the province, which are largely French and Catholic, the language used in the schools is French. Roman Catholic children are usually taught in French, and Protestant children in English. Montreal has 227 Roman Catholic public schools with more than 100,000 pupils, and 47 Protestant schools with about 30,000 pupils. But each language is taught in the schools of both types. Because English is so widely spoken on this continent, however, French children usually adapt themselves to English better than the English do to French.

High schools, classical colleges and even the universities are similarly divided along linguistic or religious lines. In certain technical and trade schools, however, both French- and English-speaking students are frequently taught in the same institution.

In Protestant high schools the course of study is much the same as that in any of the English-speaking provinces; and graduates are prepared for McGill University, Bishop's University and other institutions of higher learning. Graduates of the Roman Catholic high schools may enter either Laval or Montreal University or a normal school. A colorful sight years ago was the Laval uniform for boys—blue coat piped with white and bound with green *ceintures*. For three centuries students wore this dress.

Though English is taught in the French schools, and vice versa, recent census returns indicate that more than 70 per cent of the total French population speak only French. Yet the majority of city-born and educated French Canadians understand English and use it for everyday business in office, shop or factory. On the other hand, a large percentage of English-speaking people, even in the large cities, neither understand French movies nor use the French language in shopping or for business.

Montreal publishes daily and weekly newspapers in both languages, and these are on a par with newspapers in most

JACKSTRAWS FOR A GIANT
These logs, cut to uniform size, are raw material for Quebec's vast paper industry.

NATIONAL FILM BOARD

SOUS-LE-CAP STREET, Quebec, is one of the quaintest thoroughfares imaginable. It is a narrow, winding, old alley near the water in the "lower town" of Quebec, where French is spoken almost exclusively. There are several of these bridges connecting houses on opposite sides of the street, and the atmosphere is distinctly that of the Old World.

COPYRIGHT DETROIT PHOTOGRAPHIC CO.

TADOUSSAC, though it remains only a village, is the oldest continuously occupied settlement in Canada. It is situated at the point where the Saguenay River flows into the St. Lawrence. The Indians told Jacques Cartier strange and terrifying stories of the gloomy Saguenay. Nevertheless, in 1599, a trading-post was established, and the settlement continued in the face of famine, war and Indian attack. The little town was also long a center of Récollet and Jesuit missionary effort in the difficult task of converting the Indians of New France.

other American cities. Private radio stations broadcast in either French or English, or both, and Radio Canada in Montreal is consistently bilingual.

Montreal is the great metropolis of the province. Though located a thousand miles inland from the sea, it is one of the most important ports in the world. For a time in the nineteenth century, it was the capital of the United Province of Canada. No longer the political capital, it is still the commercial heart of the nation. Enormous grain elevators, wharves and docks store and ship more grain than those of any other port in the world. In Montreal East great oil tanks and refineries have been built where millions of gallons of crude oil are processed daily.

Like Manhattan, Montreal is an island city. It is about 30 miles long and 7 to 10 miles across. Here was the Indian encampment of Hochelaga, which Jacques Cartier discovered on his second voyage to Canada in 1535. From the commanding heights of Mount Royal, for which Montreal is named, Cartier marveled at the windings of the mighty river which had brought his frail sailing ships so far inland from the sea and promised to lead him still farther toward the golden west.

Mount Royal remains the leading landmark of the city, which has grown from an important fur-trading station in the days of New France into the financial center and metropolis of Canada. Gradually the growing city has climbed up the sides of the mountain to higher and higher levels, but the view from the summit is still worth the effort. The only way to reach the top is to walk, ride up in an open carriage or go part way by automobile or street car, for no motor cars are allowed on the final stretch of the winding road leading to the Lookout.

Downtown Montreal is a great center of shipping, by land and by sea. From the time the last ice goes out in springtime at the Strait of Belle Isle to admit the first ship of the season, the water front is filled with passenger ships and freighters from many countries. Last to clear port in the fall, before ice closes the river to navigation, are the grain ships heavily laden with Canadian wheat, many bound for Europe.

NATIONAL FILM BOARD

PATTERN OF POWER
Part of a plant that harnesses water power at Beauharnois, up river from Montreal.

The capital of the province is its namesake city of Quebec. For 150 years Quebec was the capital of France in America, and it remained the capital of Canada for 30 years after New France passed under British rule. This "city on a rock" is the most picturesque city in North America and is the only walled city on the continent. Quebec lies north of the river at the foot of the beautiful Laurentian Mountains, and its ancient citadel towers 350 feet above the quay.

Fringing the water front and on the lower slopes of the rocky bluff are the narrow steps, stone churches, steep-roofed houses and shops of the quaint old Lower Town. At the top of the bluff, separated from the Lower Town by massive walls, is the more modern Upper Town, the fashionable residential quarter. Here are the fine buildings of the provincial parliament and government departments, as well as spacious parks, fine old churches and convents. The Basilica, or French Cathedral, is famous. Begun in 1647, and destroyed by fire in 1759 and 1922, it has since been rebuilt. In the nave of the Ursuline Convent chapel are enshrined the remains of the Marquis de Montcalm. The Seminary of Quebec has buildings dating from the late seventeenth century, and, with Laval University, is the oldest seat of French learning and culture in the New World.

Dominating the Upper Town is the magnificent Château Frontenac, one of the most famous hotels in America. Built to resemble a medieval French castle, it is named for the great Governor of New France. Near the hotel is Dufferin Terrace, a famous promenade 1,800 feet long and 600 feet wide. Here hundreds of people stroll on summer evenings or coast on winter afternoons. The view from the terrace is magnificent. Directly below are the jumbled roofs of the old town, and beyond them flows the broad St. Lawrence. The St. Charles River empties into the St. Lawrence here and, in midstream, lies the island of Orléans.

The city of Levis is situated on the bank of the St. Lawrence opposite to Quebec, and on the east the Beauport shore stretches as far as Cap Tourmente, while the Laurentian Mountains extend to the east, west and north. Along the bank of the St. Charles River and in the new industrial center of St. Malo are numerous thriving factories operated by electric power from Shawinigan Falls. Quebec is the tourist center of eastern Canada, and, next to Montreal, is the largest and most prosperous city of the province.

When Samuel de Champlain first saw this incomparable site in 1608, he recognized it at once as the ideal location for a colony. Here on the north bank of the river he built his habitation of logs, fortified by a moat and stockade, not far from the present Levis ferry slip. That was twelve years before the Pilgrims landed at Plymouth. It marked the first permanent settlement by white men on Canadian soil. The Company of New France, under Champlain's leadership, was a group of Frenchmen to whom their King granted the right to colonize the new land, control its fur trade and convert the Indians.

As long as Champlain lived, his faith and enthusiasm held the pioneers together in spite of raids by the hostile Iroquois, whom he had unfortunately antagonized. After his death, in 1635, the Company of

ARMOUR LANDRY, MONTREAL

STOREHOUSE OF ENERGY ON THE SAGUENAY RIVER

Generating power house at Shipshaw, center of a thriving aluminum industry. It requires tremendous amounts of electricity, which is produced by the rushing river waters.

QUEBEC ORCHARDS have won a reputation for the quality of their fruit. The famous MacIntosh Red apple has been brought to its highest development by the growers of the Rougemont section of the Province. More of the orchards are in the Eastern Townships than in any other section but perhaps the finest one of all is in the Ottawa Valley.

AFTER SCHOOL, BOYS WITH THEIR DOG-CARTS run up and down the hills along the St. Lawrence shore, often with baby brothers or sisters for passengers. Little François or Jacques or Marie calls out "B'jour" politely to the passer-by, and sometimes a baby voice varies the greeting by making it "Hil'loa." These smiling young chaps have dressed up their pets with sun-glasses and caps; dogs and boys are equally ready to run an errand for a grownup or to hold a chariot race. French-Canadian children have a happy, though well disciplined home life

QUEBEC, ONCE NEW FRANCE

WHAT WILL THE FIGURE BE?
The gentle art of wood carving came to Quebec with the first settlers and still flourishes.

New France was disbanded. Seven years later, the Sieur de Maisonneuve founded Ville Marie de l'Isle de Montréal, and Montreal later outstripped the older colony, though Quebec remained the capital of New France.

The other principal cities of the Province of Quebec are Sherbrooke, a manufacturing and distributing center called the "garden of the Eastern Townships"; Three Rivers, one of the greatest centers of pulp and paper manufacture in the world; Hull and Valleyfield, both well known for their textile mills; St. Hyacinthe, famed for the manufacture of organs as well as for boots and shoes; and Drummondville, noted for its celanese, or artificial silk, industry. Westmount, Outremont and Verdun are large suburbs of the rapidly expanding city of Montreal.

The province supports some 12,000 manufacturing enterprises and produces annually goods worth $4,000,000,000. These manufactures run from pins to Die-

AN OPEN-AIR ARTIST ON THE ISLAND OF ORLEANS
Many Quebec farm wives are gifted at making the gay kinds of hooked rugs you see here. They may take their subjects from the life about them or design bright flower patterns.

NATIONAL FILM BOARD

FISHING FLEET OFF L'ANSE-A-BEAUFILS

The hamlet of L'Anse-á-Beaufils is on the coast of the Gaspé Peninsula, home of sturdy fisherfolk. In frail-looking craft such as these, the Gaspé fishermen put far out to sea for catches of cod.

sel locomotives for streamlined trains.

Lumber, mining and industry have attracted immigrants from many European countries to the province. Many of the newcomers are Italians, but there is a liberal sprinkling of Bulgarians, Greeks, Poles, Rumanians, Lithuanians, Czechoslovakians, Germans, Finns and Russians.

Quebec's forests cover a huge area, and pulp and paper making is the most valuable industry. The best timber in the province is red pine and spruce. Instead of selling the forest land outright to private owners, the government licenses companies to cut the timber. The whole island of Anticosti in the Gulf of St. Lawrence is leased in this way. To patrol the forests and give warning of fires, the government engages a large staff of fire rangers, uses many aircraft and maintains a large research division.

In water-power development, as in lumbering, Quebec heads the list of Canadian provinces, its total installation being 6,-173,597 horse-power of which 1,200,000 horse-power is at Shipshaw. Another 10,-000,000 horse-power is in reserve. Among the rivers harnessed for industrial use are the St. Lawrence, Richelieu, Rivière des Prairies, St. Maurice, Gatineau, Lièvre, Montmorency, St. Francis, Shipshaw, Outardes and Manicouagan. Undeveloped water-power sites are even more numerous.

In the rich minerals hundreds of feet below ground level lies a great source of wealth. These resources have created a number of new cities in northern Quebec, and prospectors are searching for further discoveries. Gold, copper, zinc and asbestos are the most valuable of the mining products. The largest single source of titanium in the world has been discovered at Lake Allard. Development is under way of the huge deposits of hematite (an important iron ore) on the Quebec-Labrador boundary.

Native arts and crafts still flourish in the province of Quebec, and many excellent examples of them may be seen in the Quebec Museum. Among the early immigrants to New France were skillful artisans who continued in the New World to pass on their crafts from master to apprentice, as had been done in Europe for hundreds of years. Early wood carvers decorated their community churches with figures and ornaments on altars, pulpits

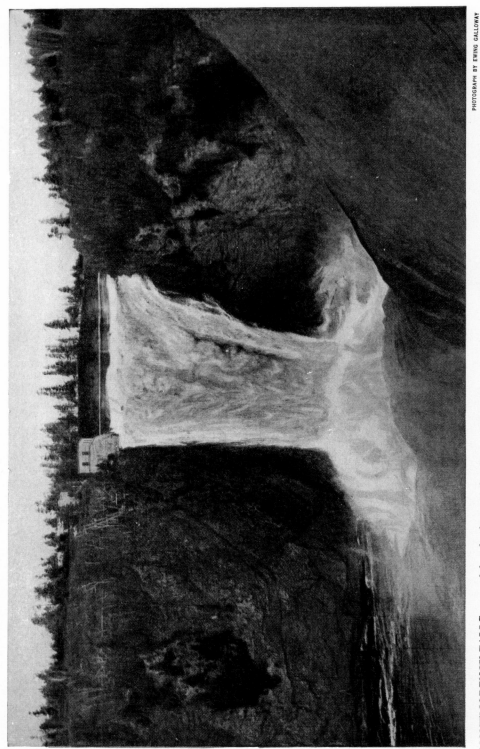

PHOTOGRAPH BY EWING GALLOWAY

MONTMORENCY FALLS, named for the famous churchman, François de Montmorency-Laval, the first Bishop of Quebec, have been famous for three centuries. They are situated in the river of the same name just before it flows into the St. Lawrence a few miles below Quebec City. The drop is about 265 feet and the greatest width about 150 feet. Much water has been diverted to generate electricity for Quebec City and the falls are no longer so imposing as formerly. Wolfe made an unsuccessful attack on Quebec from the village of Montmorency.

CAPE TRINITY on the Saguenay River, about thirty miles from its junction with the St. Lawrence, is a cliff which rises in three steps from the river to a height of about 1,600 feet. Opposite is Cape Eternity, even higher. The river itself is really a fjord, deepened by ice long ago, and is several hundred feet deeper than the St. Lawrence.

SUPPLIES FOR UNGAVA
Ungava, once a wilderness, is becoming one of North America's chief sources of iron ore.

and chancel ceilings. At Baie St. Paul and St. Jean Port-Joli, gifted wood carvers today design quaint figures of the habitants. Weaving and rug-making are also important provincial arts, and Hudson's Bay blankets are famous.

Facing the City Hall in Quebec, near the site where the first farmhouse stood, stands a splendid stone figure of Louis Hébert, the first farmer of New France, designed by Philippe Hébert, one of his descendants. An even more famous monument stands in the Governor's Garden, commemorating the British General Wolfe and the French General Montcalm, who were both killed in 1759 as a result of the decisive battle for New France on the Plains of Abraham. Though Wolfe and Montcalm were enemies, each was a gallant soldier; and both are honored in the history of Quebec. When the British won, they brought to an end the efforts of France to establish a great colonial empire in North America.

The residents of New France continued to occupy the land they loved after France yielded, and the new government helped them to adjust themselves and to maintain their homes by its recognition of their cultural and religious traditions. Though English settlers came to the new country in increasing numbers, especially after the American Revolution, the French remained the great majority in Quebec. Difficulties have naturally arisen from time to time between the two peoples, yet French and English have come to know and to respect each other more and more as the years have passed. Today Quebec is a striking example of two peoples living together in friendship as fellow citizens of one great country.

By W. P. Percival

QUEBEC: FACTS AND FIGURES

THE PROVINCE

It also includes the island of Anticosti, Bird Islands and Magdalen Islands in the Gulf of St. Lawrence. The total area (as amended by the Labrador Boundary Award) is 594,860 square miles (including Ungava, 351,780); land area, 523,860 square miles. Total population (1951 census), 4,010,235.

GOVERNMENT

Administered by a lieutenant-governor, appointed by the Federal Government, a responsible ministry, a Legislative Council of 24 members (appointed for life) and an elected Legislative Assembly of 92 members. Representation in Canadian Senate, 24; in House of Commons, 73.

COMMERCE AND INDUSTRIES

Agriculture is the basic industry; wheat, oats, barley, rye, peas, buckwheat, tobacco, mixed grains, flaxseed, corn, potatoes, turnips, hay, clover and alfalfa are grown. Fox-farming and fishing are important occupations and the province leads Canada in pulpwood production. The chief minerals are asbestos (about 75% of the world's supply), gold, copper, zinc, silver, feldspar, graphite, magnesite, mica, molybdenite, phosphate, hematite and lead. The leading industries are pulp and paper, nonferrous metal smelting and refining, cotton and its products, clothing, cigars and cigarettes, dairy products, flour and its products, railway rolling stock, synthetic textiles and silk, boots and shoes, furniture, sawmills, electric light and power, feeds, meat-packing and breweries.

COMMUNICATIONS

Railway mileage in 1949—5,122; number of telephones—604,715; in 1949—40,922 miles of roads.

RELIGION AND EDUCATION

Population about six-sevenths Roman Catholic. The schools are sectarian; in 1949, there were 9,592 schools with 638,688 pupils in attendance. There are 5 universities.

CITIES

Populations, 1951 census: Quebec, capital, 161,439; Montreal, 1,022,703; Verdun, 76,454; Sherbrooke, 49,737; Three Rivers, 45,708; Hull, 43,204.

ONTARIO, FORMERLY UPPER CANADA
The Central Province of the Federation

Ontario was settled much later than Quebec and the other provinces to the east. Until after the American Revolution there were only a few scattered settlements in what came to be called Upper Canada. Many Loyalists found refuge in the wilderness, and were soon followed by immigrants from the British Isles. Today the province is the most populous one in Canada and leads in industrial development. However, in New Ontario, the territory added to the province in 1912, there is much wilderness which has been seen by few white men. It is rich in minerals and forest wealth and is now in the beginning of a process of rapid development of its resources.

SINCE 1912 the northern boundary of Ontario has been Hudson Bay. Another water-line forms the boundary to the south—from the Lake of the Woods and Rainy River through the Great Lakes, Superior, Huron, Erie and Ontario. These latter bodies of water exert a great influence on the climate of Ontario, the "central province" as it is sometimes called, modifying both winter and summer temperatures.

The area of Ontario, 412,582 square miles (including water area), is made up of two different sections, the northern, newly opened area containing about 330,000 square miles and the southern older-settled region, about 77,000 square miles. It is in this southern region, around the lakes and on the St. Lawrence, that nine-tenths of the people live. All the larger cities and most of the manufacturing and farming districts belong to Old Ontario, but New Ontario is in the process of settlement, its forests are being cleared, its fertile belts are yielding to plow and furrow, its rocks are revealing mineral riches.

Across the surface of the province runs a ridge of Archæan rocks forming a watershed which turns some rivers north to Hudson Bay and others south to the St. Lawrence system. In general the surface is undulating, veined by rivers and small lakes, whose infinite number, especially on the northern slope, has gained the epithet "land of little lakes." Lake Nipigon (70 miles by 50 miles) is a beautiful body of water some 800 feet above sea level which may be considered as forming the headwaters of the St. Lawrence. The north shore of Lake Superior is bold and stern and sparsely settled by fisher people. From its southeast end St. Mary's River leads into Lake Huron by the falls known as Sault Ste. Marie where canals with large locks permit vessels of 10,000 tons to pass from lake to lake. The northeastern shore of Lake Huron is broken by large inlets and fringed into innumerable islands. Most of these are small but Manitoulin, the largest, is 80 miles long and 30 broad. Into Georgian Bay flows the French River from Lake Nipissing, and the Severn from Lake Simcoe.

Both Lake Superior and Lake Huron are very deep but St. Clair, the next in the series, is shallow and marshy, and Lake Erie is not very deep. From Lake Erie the water is carried over the Niagara escarpment and down the precipitous gorge into Lake Ontario, which is only 247 feet above sea level. Shipping from Lake Erie enters the Welland Canal at Port Colborne and emerges into Lake Ontario at Dalhousie.

Below Kingston, at the lower end of the lake the St. Lawrence flows through the intricacies of the Thousand Isles, then over several rapids to Montreal, head of ocean navigation. Ship canals afford passage for boats of considerable draught around the rapids. Next in size of the south-flowing rivers is the Ottawa River, navigable in parts, and the Trent. To Hudson Bay flow the Albany, Moose, Missinaibi and Abitibi—but they are not commercially navigable.

VIEW OF CROSS LAKE, COBALT, NEAR THE OLD O'BRIEN MINE

COURTESY NATURAL RESOURCES INTELLIGENCE SERVICE

THE COBALT RIVER, in northern Ontario, sprang into prominence almost in a day in 1903 when rich deposits of silver and other metals were discovered. By 1911, when the output reached its peak, 5,638 people lived in the town of Cobalt. The mines were reopened in 1946 for the production of silver and cobalt. Below, peach orchards in bloom near Grimsby.

A NARROW CHANNEL IN THE LAKE OF BAYS DISTRICT

PHOTOGRAPHS FROM EWING GALLOWAY

VACATION DAYS in Canada are full of delight. The Lake of Bays is separated by only a mile from a group of lakes almost equally beautiful, including Peninsular, Fairy, Marys, and Lake Vernon. Those residents of Toronto who cannot go so far from home in summer find delightful bathing in the clear waters of Lake Ontario at Sunnyside Beach.

ONTARIO HYDRO

A HYDROELECTRIC PLANT ON THE OTTAWA RIVER

The Chenaux Hydro-Electric Station on the Ottawa River can generate 160,000 horsepower. Opened in November 1950, it is one of four power plants built by the Ontario Hydro-Electric Company on the Ottawa. These plants furnish light and power for a large area of Canada and are helping the country in the rapid development of its abundant natural resources.

To the east of Georgian Bay lie the Highlands of Ontario, before the white man the hunting grounds of the Hurons. Here no less than eight hundred waterways, lakes, rivers, streams are to be found. Beautiful Muskoka with its three lakes, Muskoka, Rosseau and Joseph, is a great playground today for the city-dwellers of western Ontario. "Islands abound, from a tiny one-tree speck of earth or a bare cone of rock, to a thousand-acre isle stranded mid-lake in beautiful Rosseau. Each turn of steamer or canoe reveals a new vista." Algonquin Provincial Park, to the east of Muskoka, occupies over a million acres of forest, game and fish preserve. Moose, deer and beaver are rapidly increasing under the protection of the Government on the wooded shores of lake and stream.

Temagami ("lake of deep waters"), another reserve, offers a panorama of yet hidden beauty. The main lake contains over 1,400 islands and covers an area of 90 square miles. Caribou and moose abound in its forests; its waters teem with bass and trout and pickerel. During the construction of the Timiskaming and Northern Ontario Railway (built to tap timber areas and serve small farming communities) the discovery of the great silver mine at Cobalt was made. To the west is Sudbury, famous for its rich nickel and copper deposits.

New Ontario or Northern Ontario was formerly a part of the territory of Keewatin but in 1912 the Federal Parliament extended the boundaries of Manitoba and Ontario north to Hudson Bay and gave eight great districts to Ontario. Four times as large as Old Ontario, it is a region of forests, rocks, rivers, lakes. While the mineral areas are beginning to yield richly, much is still unexplored. Nipissing, named for its great lake, is a sportsman's paradise, and contains Algonquin Park and part of Temagami. Timiskaming, stretching to Hudson Bay, was the hunting ground of the earlier French and English fur trappers. Prospectors of a later day discovered the rich deposits at Cobalt, Elk Lake City and Gowganda, and railway scouts prospecting for good lands for settlement came upon the immense stretch of fertile country known as the great Clay Belt, stretching through Algoma and Timiskaming. Algoma, a vast territory extending 360 miles north from the Soo to the Albany River, is a true land of lakes and rivers

DEPARTMENT OF HIGHWAYS, TORONTO

A CLOVERLEAF INTERSECTION ON THE QUEEN ELIZABETH WAY

The Queen Elizabeth Way, the four-lane expressway from Toronto to Niagara Falls, forms a cloverleaf intersection outside Toronto with Highway No. 10, another four-lane road. The loops and ramps are designed so that a motorist may enter or leave either road safely without having to cross a lane where there is traffic moving in the opposite direction.

In its 200 miles of coast-land on Lakes Huron and Superior the largest town is the historic little Sault Ste. Marie named by French *voyageurs* of the seventeenth century from the falls in the St. Mary's River. Thunder Bay has a grim frontage on Lake Superior and a forested region lake-studded to the north. Fort William and Port Arthur, twin cities with immense docks and elevators stand to-day where formerly stood the rude lodges of the North-West Company of fur-traders.

Rainy River, so called from the perpendicular fall between lake and river which gave forth spray like rain, was discovered by the French and ministered to by the Jesuits. Kenora, an early Hudson's Bay Company's fort under the name of Rat Portage, is the boundary district between Ontario and Manitoba and has been the site of boundary disputes. To-day its many lakes form a playground for the citizens of Winnipeg. Patricia, eighth and last district of New Ontario, is the largest of all, adding fifty-six per cent to the area of Ontario. It has a shoreline of six hundred miles on the James and Hudson bays, but as yet has not been fully explored.

But the history of New Ontario is still in the making, while that of Old Ontario has inscribed many a crowded page in its brief 160 years of life. The wooded wilderness lying to the south of the Laurentian rocks was known only by the fur-trapper and the explorer at the end of the eighteenth century. Champlain had penetrated from Quebec as far as Georgian Bay, the Récollets and the Jesuits had labored for the Hurons, LaSalle had seen the country as he made toward the Mississippi; but the Hurons had been exterminated and the English fur-trappers had only a few fortified posts on Hudson Bay, and the French at Fort Frontenac and Michilimackimac and Sault Ste. Marie. When the French lost Quebec in 1763 what was known as "Upper Canada" lay a wilderness for twenty years. In 1782 bands of United Empire Loyalists began to come in. Surveyors sent by General Haldimand chose lands in four districts for the loyal exiles: along the St. Lawrence opposite Fort Oswegatchie, around the Bay of Quinte above Fort Cataraqui, in the Niagara Peninsula opposite Fort Niagara, and in the southwest section within reach of Fort Detroit. The settlers were of varied origin—Highland Scots, German, Dutch, Irish, Eng-

YOUNG CATTLE ON PASTURE ON A FARM NEAR OTTAWA

COURTESY NATURAL RESOURCES INTELLIGENCE SERVICE

ONTARIO is an empire in itself with many differences in elevation, soil and climate. In extent it is as large as France and Germany with Massachusetts and Connecticut thrown in for good measure. In the upper picture are contented cattle in the dairy region near Ottawa, while the lower picture shows a log-jam in the rocky bed of the Montreal River.

VINEYARD AND ORCHARDS AT GRIMSBY

COURTESY NATURAL RESOURCES INTELLIGENCE SERVICE

GRIMSBY on the Southern shore of Lake Ontario is in the very heart of the grape-growing section, and everywhere one may see the long straight rows of vines twined upon wire trellises. A dozen varieties are grown, though the Concord is the most popular. The lower picture shows a view of the Ottawa River with the low Laurentians in the distance.

lish. Most of them had already gone through the pioneer stage in settling new lands and brought valuable experience to bear upon their problems. The government allotted them lands, gave them implements, seed-grain and at first even food. To the forest succeeded small cleared areas, which in turn bore grain and food for man and beast. As well as lumberman and farmer, the settler of those days must be a trapper also to supply his family with food and clothing. Slowly the settlements grew and the trails widened, though the waterways long continued the chief avenues of communication.

Until the end of the century the Loyalists continued to come, and there were besides the loyal Indians of the Six Nations. For these land was purchased, a tract six miles wide on each side of the Grand River in western Ontario, and here under the Mohawk chief, Joseph Brant, many settled.

To thirty years of struggle with the wilderness succeeded a struggle with their neighbors across the line. In the War of 1812, the Loyalists bore heroic part, fighting as they were to defend the homes so hardly won. When the war began the population of Upper Canada numbered about eighty thousand, for Simcoe the first governor had done all in his power to encourage immigration; and many besides the Loyalists had come in from the United States.

Three years' fighting was a serious setback to the work in field and homestead but the pioneer women were cast in heroic mold and the work was not stayed. At the close of the Napoleonic Wars many British veterans began to pour into Upper Canada and were given lands in townships to the rear of those

NATIONAL FILM BOARD

WHERE PEACETIME USES OF ATOMIC ENERGY ARE EXPLORED

At Chalk River, Ontario, the Canadian Government maintains a pilot plant that produces materials for the release of atomic power. No war weapons are produced here; scientists are studying possible ways in which the atom's energy can make man's life easier and better. Workers at the plant live in the near-by town of Deep River, built for them by the Government.

LABORATORY TECHNICIANS at the government's Polymer Corporation near Sarnia, Ontario, test a sample of buna-S rubber. Canadian synthetic rubber must meet rigid standards.

settled by the Loyalists, or in unoccupied ones lying between. By 1826 the population had increased to 166,000; by 1836 it was 374,000 and in 1841 it was 456,000. People lived on their own land, towns were comparatively small. Kingston was the largest; then came York (later Toronto), London, Hamilton, Brockville.

The British settlers brought in good livestock and a knowledge of breeding which placed agriculture on a firmer basis. Oxen were as yet more numerous than horses for they were hardier. But now to skins were added homespun garments from the wool of the sheep, and coarse linen fabrics from the homegrown flax. Roads pierced the forests and broke down the isolation of frontier settlements. Mills, schools, churches acted as magnets to draw people together. The Loyalists had sacrificed their first homes on the altar of freedom; they were not content until the Constitutional Act of 1791 separated their province from Quebec or Lower Canada, and gave them English civil and criminal law, a legislative assembly and council and a lieutenant-governor. The Ottawa River was chosen as a boundary, but the two seigniories of New Longueil and Vaudreuil were still kept by Quebec although on the western side of the river.

In 1841 the two provinces were united and given responsible government. For twenty-seven years the neighbors were yoked together but the equality of representation granted to them became unfair to Upper Canada as her population first equaled and then surpassed that of Lower Canada. Separate schools conceded to the Roman Catholics in 1863 contributed another grievance. When federation of the provinces was mooted Upper Canada was strongly in favor of

PHOTOGRAPH FROM EWING GALLOWAY

MAIN CHANNEL at Honey Harbour, Ontario, is not very broad, but it is deep enough for steamers to pass through allowing the passengers a close view of the delightful scenery along its shores. From Honey Harbour, one may take a boat to Beausoleil Island, largest of the thirty Georgian Bay Islands which form the Georgian Bay Islands National Park. Beausoleil Island as well as much of this region is ideal for recreation such as boating, fishing, swimming and hiking. Campsites have been built and hotels serve vacationers as can be seen here.

ST. MARY'S RIVER connects Lake Superior with Lake Huron, and also separates Canada and the United States. The passage, however, is impeded by dangerous rapids—Sault Ste. Marie. To avoid them, Canada and the United States built the "Soo" canals with four locks on the United States side of the river and one on the Canadian side. The canal is free to vessels of either nation. Here a vessel is in the lock of the canal on the Canadian side. The traffic through these canals is much greater than that passing through even such a waterway as the Suez Canal.

ONTARIO, FORMERLY UPPER CANADA

it, for it meant her freedom again from her uneasy yoke fellow. Since Federation (1867) she has been known as Ontario, and Lower Canada as Quebec.

The racial origin of the present population of Ontario is reported by the census as predominantly of the "British Races," chiefly English, Irish and Scotch in the order named. There were over eight hundred thousand of "other" European origin, chiefly French and German, though almost every people in Europe is represented. There are a few Asiatics, and over thirty thousand Indians.

The Cities of Ontario

Great cities and flourishing towns have sprung up in Ontario. There is one city of over half a million—Toronto—two others larger than 200,000—Hamilton and Ottawa—and two more over 90,000—Windsor and London. We tell about these and other Canadian cities in the chapter that begins on page 121.

After the American Civil War large tracts of land were opened to farmers in the Middle West, and this not only attracted men from Ontario but the better conditions for growing wheat destroyed the crop in the older east. Then with the development of the Canadian Northwest—the real wheat belt—consequent upon the completion of the Canadian Pacific Railway in 1885, farmers in Ontario suffered from genuine depression. Readjustment of crops took time, but the process developed the fruit-growing and dairy-farming to a very considerable degree.

Ontario today is a leading fruit-growing province of the Dominion. She has an abundant rainfall, a suitable soil, plenty of warm sunshine. The Prairie Provinces to the west, the United States to the south, Europe to the east furnish good markets which her facilities by rail, canal and river can easily supply.

Where the Fruit Grows

The St. Lawrence fruit belt extends from the eastern end of the province to the city of Kingston and grows many of Canada's famous apples. In the Ottawa Valley between l'Original and Pembroke, the Yellow Transparent, Duchess, Wealthy and McIntosh flourish. Prince Edward County on the Bay of Quinte is a notable fruit area, and the orchards continue west through the counties of Northumberland, Durham and Ontario. Toronto provides a great market for central Ontario.

At Port Credit a new small fruit and vegetable country begins stretching along the lake shore to Hamilton. From the base of the ridge or mountain at Hamilton, a great fruit market, a level floor runs to the shore of Lake Ontario. This floor—once a lake bed—forms the far-famed peach belt of Ontario. Grapes grow here too in profusion—the Concord, Worden, Champion, Niagara and Delaware. The sweet cherry is cultivated only in the Niagara region, though the sour cherry is widely grown as well over that part of the province west of Toronto to Georgian Bay. The Lake Erie district is a home of successful fruit-growing, and in the Georgian Bay fruit belt large crops of winter apples, plums and pears are shipped from Owen Sound.

Great Mineral Production

Ontario leads all the other provinces in mineral production. Over the ten-year period ending in 1950, the average worth of the annual mineral output was $260,870,000. Ontario's output, approaching a value of $400,000,000 a year, now accounts for a third of Canada's total mineral production. Though northern Ontario has not yet been fully explored, what little is known gives zest to further development. There is no richer mineralized area than that of Sudbury to the north of Superior with its vast known reserves of base and precious metals in the ore bodies of the nickel ridge and the copper-zinc-lead deposits of the basin which the ridge encloses.

Farther north is the famous Cobalt camp where the silver-cobalt mines for many years turned out some of the most profitable ores the world has seen. Farther north still are the gold mines of Porcupine which has produced

A SIPHON holds the close attention of the boys in a physics class of a public school in Ottawa. The lesson is a step on the path to the marvels of twentieth-century science.

gold worth hundreds of millions and is still one of the largest gold-producing camps in Canada. The Hollinger Mine has been the largest producer of the group. The younger camp at Kirkland Lake has now surpassed the older, and the Lake Shore Mine has in some years produced more than any other.

The distribution of this wealth affects all classes of the community. Mining is even more important than agriculture in providing freight for railways. About one-third of Canada's railway tonnage is made up of the produce of the mines. Many of the virgin areas which are now under exploration for minerals will eventually be taken up for farming purposes, for there are immense stretches of arable land in the north which could sustain a large population apart from mining industries. On the bank of the Grand River in southwestern Ontario is a gypsum mine which has been worked steadily for more than a hundred years; in south-western Ontario petroleum also has been produced for many years and natural gas is used there for cooking and heating.

In the north of Ontario there is a large tract of forests. The Georgian Bay district contains the largest area of white pine in the world and sufficient to supply the trade for a number of years. Ontario has a considerable amount of hardwood, and an inestimable supply of spruce. In the north the characteristic trees are the maple, beech, birch, elm, ash, oak, hickory, pine, cedar, spruce, hemlock. The forest growth of south Ontario is different and the predominant trees are the oak, hickory, chestnut, buttonwood and tulip. The provincial government does much to encourage forestry, by schools, by nurseries, by fire protection, by reforestation.

Another Loyalist settlement in the Niagara Peninsula is but a short boat's journey from Toronto. After crossing

GOVERNMENT DRIVEWAY which circles the entire city of Ottawa is only a part of an elaborate system of boulevards thirty miles in length. For the most part these drives are bordered with flowers and shrubs and make motoring a delight. Ottawa is an attractive city as well as a busy one. Long ago it was called Bytown from Colonel By who made the surveys for the Rideau Canal. It was chosen as the capital of Canada by Queen Victoria when Montreal, Quebec, Toronto and Kingston were all vigorously expressing their claims for that honor.

PHOTOGRAPH FROM EWING GALLOWAY

THE PEACH ORCHARDS of the Niagara Peninsula are famous. This stretch of land between Lake Erie and Lake Ontario is practically an immense orchard and vineyard. Apples, peaches, pears, plums, cherries and other fruits grow to perfection and there are miles of vineyards as well. Though fruit has been grown in this section since it was first settled, commercial orchards and vineyards have developed only with improved railway facilities. Peach-growing upon a commercial scale began about 1890 and in 1904 the first fruit was sent to Winnipeg.

ONTARIO, FORMERLY UPPER CANADA

Lake Ontario and entering the Niagara River the steamer passes by densely wooded banks, where stands Niagara-on-the-Lake, formerly Newark, the first capital of Upper Canada. Here again, as in Kingston, Loyalists succeeded Indians and Frenchmen. Farther up the Niagara River rise the Queenston Heights where in 1812 was fought a memorable battle. Here in heroic resistance fell the Canadian leader Sir Isaac Brock and in his honor a tall monument rises to-day. Either at Queenston or Lewiston it is necessary to land and go by rail up the precipitous gorge toward the thundering Falls of Niagara. To the west of the falls the large station of the Hydro-Electric Commission stands to gather in the power that is thence transmitted hundreds of miles through the province.

Through the garden of Ontario from Niagara to Hamilton, orchards and vineyards stretch their ranks, spreading in spring a lovely panoply of pink and white and in autumn their luscious wealth of fruits. St. Catherine's, high-set upon its hill, has healing waters in its wells, and a fine boys' school. Hamilton, at the head of Lake Ontario, has many industries, which include the manufacture of guns, buttons, brass and jewelry. It stands moreover on the highway of traffic into western Ontario, and between Detroit and Buffalo factories. By an inclined railway one climbs its "Mountain" to view the far-reaching panorama on every side of fruit lands and tree-encircled farmhouses. It is a scene worthy to be remembered.

On the Grand River where once the Indians forded, stands the modern city of Brantford. Not far away is the burial place of Joseph Brant—Thayendenaga—famous Indian chief. St. Thomas, one of the most important railway centres of the country, and London on the Thames with its Western University, are situated in the fertile stretch of country bordering on Lake Erie. Sarnia on Lake Huron is a starting point of navigation on the upper Great Lakes; the little town of Sault Ste. Marie on the canal between Huron and Superior has iron and steel works, pulp and water mills, dry-docks and shipbuilding plant. Port Arthur, at the head of Canada's inland water route 1,217 miles from Montreal, has a large number of lumber companies operating, as well as water power plants, a pulp and paper mill and various other industries. Near by the city of Fort William has great grain elevators, and the two cities form a connecting link between the Prairie Provinces and the Atlantic.

ONTARIO: FACTS AND FIGURES

THE COUNTRY. Central province of Canada with a total area of 412,582 square miles; land area 363,282 square miles. Population (1951 census), 4,597,542; Indian population is about 30,339.

GOVERNMENT. Administered by a lieutenant governor appointed by Federal Government and a ministry responsible to a Legislative Assembly of 90 members. Representation in Canadian Senate, 24; in House of Commons, 83.

COMMERCE AND INDUSTRIES. The province is rich in agricultural and mineral resources. Chief farm products are milk, hay and clover, oats, eggs, tobacco and wheat. Valuable timber resources include spruce, pine and poplar. Leading minerals are nickel (world's principal source), gold, copper, platinum, sand and gravel, iron ore and cement. Leading industrial province with nearly 13,000 manufacturing establishments. The chief manufactures are automobiles and parts, electrical equipment, pulp and paper, nonferrous metals, prepared meats and flour.

COMMUNICATIONS. Heavy shipping on Great Lakes, 5 of them bordering Ontario, several canals and the principal rivers. Railways, 10,464 miles; highways, 73,000 miles; 50 licensed airfields; 1,260,000 telephones.

RELIGION AND EDUCATION. Population more than three-fourths Protestant; no discriminatory laws. Complete state system of elementary and secondary schools with 823,000 pupils in 1950. Five universities, several colleges and professional schools, and one military college.

CITIES. (Populations, 1951.) Toronto (capital), 1,081,460; Hamilton, 201,296; Ottawa, 195,067; Windsor, 123,849; London, 95,612.

Four Great Western Provinces
British Columbia and the Prairie Provinces

The provinces of the Far West are really two groups from a geographical standpoint. The Prairie Provinces are much alike, composed for the most part of rolling plains, but much of British Columbia is mountain, valley or plateau. Though there were settlements here much earlier, most of the growth in population is of comparatively recent date. The Prairie Provinces raise the best wheat in the world, and British Columbia has within its borders many varieties of soil and climate. Though the trapper has been followed by the rancher, the farmer, the lumberman, and the factory-worker, many thousands of pelts are still taken every year in the unsettled portions of these Provinces.

IN this chapter it is convenient to consider the Prairie Provinces—Manitoba, Saskatchewan and Alberta—as one unit, and the western province of British Columbia beyond the Rockies as another.

Three events decided the destiny of the Prairie Provinces: in 1670 Charles II issued a charter to the Hudson's Bay Company; in 1783 Montreal fur merchants combined their rivalries in the North-West Company; in 1811 Lord Selkirk received from the Hudson's Bay Company a grant of 116,000 square miles on the Red River. The two companies leaving the open plains of the south to the buffalo, went up the Saskatchewan and reached the great fur country of the Mackenzie. On the Saskatchewan their trading-posts, Fort Carlton, Fort Cumberland and Edmonton House, became famous. From Fort Chippewyan on Lake Athabasca, Alexander Mackenzie in successive journeys explored the Mackenzie to the Arctic, and the Peace River to the Pacific Coast. The publication of his Voyages in 1801 directed the attention of a young Scotch philanthropist, Lord Selkirk, to the Lake Winnipeg and Red River country as a promising field for settlement by distressed Scotch crofters. He purchased a large tract of land in what is now Manitoba, Minnesota and North Dakota, and in the spring of 1812 under Captain Miles Macdonell a party of settlers arrived from Scotland.

The establishment of the new colony was not welcomed by the North-West Company, nor indeed by the Hudson's Bay Company though the latter was not openly hostile. The Nor'Westers resented the threat to their monopoly of fur routes and when in 1814 Governor Macdonell prohibited the unlicensed export of provisions from the colony, the traders lured away many of the colonists to Upper Canada, destroyed the buildings in the settlement and drove the remaining refugees to the shores of Lake Winnipeg. Again by a fresh band of Scotch the settlement was revived. Then the Nor'Westers laid plans for more decided action. They assembled a band of Indians and halfbreeds for an attack. Hostilities broke out prematurely, and the governor, Semple, and twenty of his men were killed in an affair afterward known as the Massacre of Seven Oaks. Lord Selkirk meanwhile had secured military aid for his settlement, Swiss mercenaries chiefly from the War of 1812, and with these once more brought back his little band and re-established them. He found it impossible however to secure the punishment of anyone responsible for the Seven Oaks Massacre, though for his own resort to arms he was fined. Bitterly disappointed and dispirited Selkirk retired to England and died three years later. Nevertheless his belief in the great possibilities of the western country has been more than justified.

For fifteen years after his death the executors of his estate controlled the colony, spending large sums of money in efforts to improve agriculture and establish industries. For years the little colony met disaster after disaster: a plague of

YOHO NATIONAL PARK in British Columbia contains some wonderful waterfalls. The Takakkaw Falls, a few miles from the town of Field, jump in three successive leaps, first 150 feet, then 1,000 feet, and finally 500 feet more, tumbling at last in a cloud of spray into the Yoho River. We show here a view of the tumultuous milky green water of the middle falls.

TWIN FALLS are also in the Yoho Valley in Yoho National Park. They are, however, less accessible, as no road for motor cars has yet been constructed, and the visitor must approach from the Yoho camp either on foot or on the back of a sure-footed mountain pony. The falls are fed by the melting Yoho Glacier lying in the high mountains above.

locusts ate the land bare; the Red River flooded and swept away settlements and stock; many colonists in despair left the settlement. But the surviving courageous pioneer was at length rewarded by a series of good harvests and "Peace and Plenty" became his watchword.

Meanwhile the Nor-Westers and Hudson's Bay Company had united (1821) under the latter's name, and becoming aware of the value of the new settlement sought its control by purchasing the Assiniboia district. Difficulties arose out of this situation. The Company was, while governing the colony, actually its rival in trade. Cleavage in interest became manifest and a movement to liberate the colony gained widespread support.

The Region Transferred

Such was the state of affairs when the Canadian government in 1868 took over the territory of the Hudson's Bay Company. The Red River and other western settlements contained at this time a population of 12,000 of whom about five-sixths were half-breeds of French or Scottish blood. The government did not consult the settlers about the transfer of their territory—and seemed in their new surveys to be contemplating forfeiture of land. The Métis rose under Louis Riel and set up a provisional government but Riel's extremes roused such opposition that arrangements were finally completed for the Red River Settlement's entry into the Dominion as the Province of Manitoba in 1870. Riel fled but returned fifteen years later to the Saskatchewan and again organized a provisional government for the Indians and half-breeds. His insurrection was suppressed without serious difficulty and Riel was executed for high treason.

British Columbia Joins

The way was now opened for the extension of the Dominion to the Pacific Coast where a colony (of which we speak below) had already been established. Terms of union were settled upon wherein the government promised to commence within two years and complete within ten a railway to the Pacific Coast, and British Columbia in 1871 entered the Dominion.

After taking over the prairies the government made surveys and prepared for settlement, but for a while the tide of incoming settlers was not high, and varied from year to year according to conditions. A good crop one season, a summer frost another year, an advantageous change in land regulations, all had their effect. First-comers settled on half-wooded spots as if afraid of the winter on the open plains, but, in 1875, 6,000 Mennonites took up their abode on prairie lands along each side of the Red River. The first Ruthenians arrived in the nineties and founded large colonies in Northern Saskatchewan and Alberta. Icelanders and Scandinavians followed in the north. And with the beginning of successful grain-growing, elevators sprang up along the line and mills for grinding. Ranching was attempted in southern Alberta, the railway touched at more points, special land terms were offered by the Canadian Pacific.

The Tide of Immigration

Finally at the end of the century, Clifford Sifton, Minister of the Interior, by an intensive campaign of advertising in newspapers and pamphlets in Britain, Europe and the United States induced a great tide of settlers. The total immigration into Canada in 1897 was 21,000; in 1902 it was 67,000; in the following year 125,000 and by 1913 it reached nearly 400,000. During this period of rapid expansion nearly as many settlers came from the United States as from Britain, while just over a quarter of the total immigration came from Continental Europe. Not only did people arrive from across the seas and from the south but the people of the eastern provinces moved westward into the new lands.

With the new settlement and resultant crops further transportation seemed necessary. Sir Wilfrid Laurier's government in 1903 agreed with the Grand Trunk Railway Company for another transcontinental railway, the government itself building the part from Moncton to

STANDARD OIL CO. (N. J.)

OIL REFINERY AND LABORATORIES IN TURNER VALLEY, ALBERTA
Alberta is the heart of Canada's rich oil fields, which are making the country an important one in the world petroleum market. The Canadian oil industry first developed in Turner Valley.

THE MASSIVE RANGE is sometimes called the Bourgeau Range from the name of one of the principal peaks. The other important peaks are Pilot and Brett. Those with vivid imaginations declare that they can see on Pilot a recumbent figure which suggests the Duke of Wellington. These peaks are visible from many points. A particularly good view can be obtained from the Bow River, near Banff, and from several different points on the Banff-Windermere highway. This road forms a part of a loop which encloses much impressive and inspiring mountain scenery.

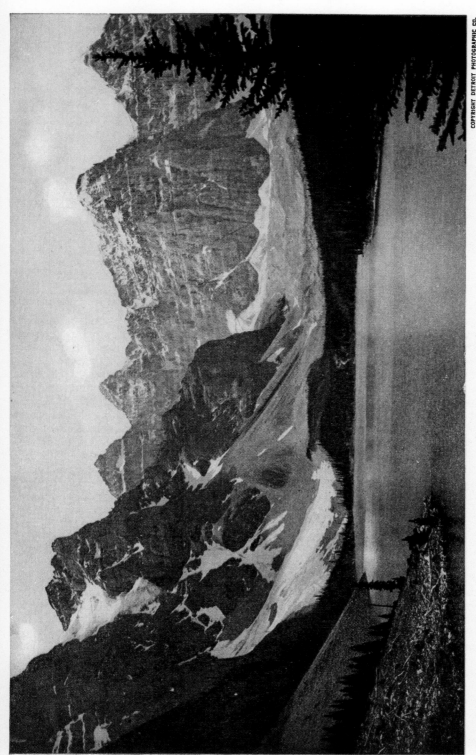

MORAINE LAKE is only nine miles from Lake Louise, and was so named by the discoverer from a morainal deposit which obstructs the outlet and which was left by the last ice-sheet. On one side of the lake, in a semi-circle, are the famous Ten Peaks, formerly numbered, though now six of them, all over 10,000 feet in height, are named. Except for a tiny tea house, there is no sign of man's presence in the region which remains in its primitive beauty and grandeur. The peaks show many marks of their struggle with the elements through the ages.

103

INTERNATIONAL HARVESTER COMPANY OF CANADA, LTD.

MODERN METHODS SPEED THE HARVESTING IN WESTERN CANADA
A caterpillar-type tractor pulls a large combine that harvests and threshes the grain while moving over the field. Canada ranks high among the great wheat-producing countries of the world.

Winnipeg. Thence the railway company went on to Edmonton and through the Yellowhead Pass to the Pacific. A branch was made from the main line to Fort William so as to give an outlet for grain by way of the Great Lakes. A third system, the Canadian Northern, under William Mackenzie and Donald Mann, by construction and purchase had in 1902 nearly 1,300 miles of road in operation between Lake Superior and Saskatchewan and by 1915 the links of a third transcontinental railway system had been welded together. Both the Grand Trunk Pacific and the Canadian Northern found the cost of construction beyond their estimates and were forced to obtain loans from the government, which in turn sought means to recover its investment. Finding none, after the war, it took over the companies altogether under the name of the Canadian National Railways.

What of the problems connected with transfer to the government, with settlement and railway development? First it was necessary to secure title to lands from the Indians. This was done by a series of wise treaties, in which the rights of the aborigines were protected. In subdividing the land for settlement, a plan similar to that in use in the United States was adopted. An area one mile square was employed as the unit of division and this area, known as a section and containing 640 acres, was divided into quarter-sections of 160 acres. Townships were of uniform size, six miles square and divided into thirty-six sections. At first settlement followed the streams and at Battleford, Prince Albert, Duck Lake, St. Albert and Edmonton colonies were formed, but with the construction of the Canadian Pacific Railway, lands along the railway became most valuable. Order was kept on the plains by the organization of that famous body, the Northwest Mounted Police. At first these police were a police of the wilderness looking after the Indians and criminal whites; as villages and towns sprang up and farmers grew crops they had to "maintain the law" in town and country. When the western towns grew larger they set up their own police forces leaving the Mounted Police as a rural constabulary. In 1905 the "Mounties" received the title of Royal from King Edward VII. During the first World War two squadrons of "Mounties" saw active service overseas. In 1920 the Royal Northwest Mounted Police were combined with another police force, the Dominion Police, to form the Royal Canadian Mounted Police.

The Force rendered yeoman service in World War II, guarding vulnerable points throughout all of Canada. It is the sole police force operating in the Yukon Territory and the Northwest Territories, and performs a variety of services in all provinces and both Territories for the Canadian Government.

When the provinces entered Confederation they received annual grants from the Canadian Government. A persistent effort was made by Manitoba to get better terms and in 1912 not only was its annual grant increased but the province was enlarged to more than double its size by the extension of the northern boundary to 60° N. and by the addition of a large triangle extending to Hudson Bay, thus giving the province a seacoast.

Sixty years ago the great plain stretching from the Red River to the Rockies, the abode of over a million and a half of people and forming the provinces of Saskatchewan and Alberta, was inhabited only by wandering bands of Indians, by herds of buffalo and a few intrepid furtrappers. Agriculture has been the chief source of the wealth of both provinces; manufacturing has steadily increased; and the coal production of Alberta surpasses that of Nova Scotia. The farmers, mainly occupied in grain-growing, early formed co-operative organizations to care for the storing and selling of their crop. Thus the wheat-pool has gradually emerged as the medium for the sale of western grain. Owning large elevators at Fort William and Vancouver the pool handles the grain from the local railway to the purchaser. The opening of the Panama Canal has made it possible to send grain to Europe throughout the year without so long a train haul as the eastern ship-

WRESTING ATOMIC FUEL FROM THE EARTH—BEAVER LODGE MINE
New proof of the wealth to be unearthed in Saskatchewan is the uranium mine (above) at Beaver Lodge. As a source of uranium-bearing ore, Canada is second only to the Belgian Congo.

COURTESY CANADIAN NATIONAL RAILWAYS

KINGSMERE RIVER is a very short stream which joins Kingsmere Lake and Wakesiu Lake in Prince Albert National Park, Saskatchewan. One may accomplish a circuit of a large part of the park by canoe, making only a few short portages between the many lakes and streams. The whole park shows many evidences of the work of the ice-sheets of the past.

MOUNT SIR DONALD VIEWED FROM MOUNT ABBOTT

COPYRIGHT DETROIT PHOTOGRAPHIC CO.

THE SELKIRK MOUNTAINS, a part of which are included in Glacier National Park, are not a part of the Rockies. Mount Sir Donald, named for Sir Donald Smith (later Lord Strathcona) is the highest, 10,808 feet. The great Illecillewaet Glacier shown in the lower picture is near the railroad. Though fed by a great snow field above, it is melting faster than it grows.

NATIONAL FILM BOARD

TRUCKLOADS OF SUGAR-BEETS ROLL INTO A CANADIAN FACTORY

Sugar-beets grow well in cool countries, and Canada's crops have increased enormously during recent years. The sugar-beet is usually larger than the table vegetable and yellow or white in color. When the sugar has been extracted, the pulp can be fed to animals. Another important product that is made from sugar-beets is molasses.

ping entails. In the year 1931, the Hudson Bay Railway was finally completed; it offers for a brief summer period a sea-haul shorter by a thousand miles.

The citizens of the Prairie Provinces are very varied in origin—and for this reason education is of vital importance. As early as 1877 the University of Manitoba was established, and it has become the center of the provincial educational system. The University of Alberta at Edmonton was established in 1905 and the University of Saskatchewan in 1907. In all of these universities agricultural education is an important side of the curriculum.

A map of Saskatchewan would be easy to draw as its north and south boundaries are parallels of latitude (49°—60° N.) and its eastern and western boundaries are meridians of longitude (109°—104° W.). The area of the province is 251,700 square miles—slightly less than that of Manitoba and greater by 5,000 square miles than the combined areas of Great Britain, Ireland and Norway. The country is for the most part open rolling prairie at an average altitude of 1,500 feet above sea level. In the north it is more broken and as yet but slightly developed. The climate is continental, the summer temperatures almost tropical but with cool nights. The winter temperature occasionally reaches 40° below zero but it is tolerable on account of the dryness and absence of high winds. Light rains fall in summer and only a moderate snow in winter. This dry vigorous climate is healthful for stock. Wonderful lakes are to be found in the northern part of the province, forest-set and rock-framed. The population (1951 census) is 831,728, about three-fifths of which is agricultural and lives on the land. Their produce constitutes a large part of the provincial income. Acreage value of field crops in Saskatchewan is enormous; the wheat, oats and barley grown are of very fine quality. The wheat is hard and heavy, the oats plump and hard. Rich grasses cover the land in parts not cultivated and upon the grazing lands vast herds of cattle and sheep feed. The only cities of any size are Regina, the capital, Saskatoon and Moosejaw, which will be mentioned elsewhere.

Lying between Saskatchewan on the east and the Rocky Mountains and the 120th meridian on the west, and bounded on the north and south by the Northwest Territories and the United States respectively, is the province of Alberta. Its area, 255,285 square miles, is the greatest of the Prairie Provinces. Formerly almost exclusively a ranching country, it has now become a great wheat-producing region, the frontier of the grain-growing area. In the southwest, considerable coal and oil-mining are carried on; lumbering is important in the more mountainous western parts and in the north, while some ranching is still pursued in the less populous sections. Rainfall is somewhat scanty in southern Alberta but extensive irrigation areas have been formed east and north of Lethbridge and along the Canadian Pacific Railway from Calgary to Medicine Hat. Central Alberta is the best settled area in the province with a rich soil and sufficient rainfall. Northern Alberta is sparsely populated, but well watered. Nearness to coastal influence and the prevalence of the chinook wind modify the climate considerably. The great divide in the Rockies forms the western boundary line of Alberta leaving much beautiful mountain scenery within her borders. The important cities are Edmonton, the capital, and Calgary, which are described elsewhere.

Manitoba, the most easterly of the Prairie Provinces and also the oldest in point of settlement, extends roughly from a line joining the west coast of Hudson Bay and the Lake of the Woods to a line approximately the 102nd meridian west. On the north and south it is bounded by the 60th and 49th parallels of latitude respectively. The total area of Manitoba is 246,512 square miles, of which a large part is rolling prairie land, the home of the buffalo in the days of Indian and fur-trader. These prairies are now wheat fields, and instead of Indian and trapper we have the farmer and manufacturer. About 775,000 people live in the province. Those of British descent make up about 50 per cent of the population. There are many of French descent in and around St. Boniface and a number of other nationalities throughout the province. About 16,000 Indians still live in Manitoba, most of them engaged in farming, herding cattle, trapping and fishing on reserves. We tell about them in another chapter. Win-

NATIONAL FILM BOARD

A SEAPLANE ARRIVES AT WEST TAHTSA CAMP ON BURNS LAKE

A small plane, equipped with pontoons, lands on Burns Lake, British Columbia. The plane is delivering mail and supplies to people whom other forms of transportation cannot reach.

COPYRIGHT DETROIT PHOTOGRAPH .C CO.

BOW FALLS on the Bow River at Banff are only fifty feet high, but nevertheless they fall in a cloud of spray. The river rises high in the mountains, and at first its course is rapid and rough. A few miles above Banff it slows down and widens out into attractive pools that reflect the many mountain peaks around the Bow valley. Just beyond the bridge the river quickens its pace and forms rapids, before taking the jump over the falls. The view of the many surrounding peaks from Bow Bridge is a favorite with the many visitors who make a stop at Banff.

COPYRIGHT DETROIT PHOTOGRAPHIC CO.

EMERALD LAKE, about seven miles from the town of Field, is also in Yoho National Park, and does not belie its name. The lake is entirely surrounded by evergreens which grow down to the very edge of the water, and there is hardly a stone visible anywhere along the shores. The waters too are emerald in color and show many shades and tints of green in varying lights. Our picture looks toward the Van Horne Range. Mount Burgess and Mount Waptu, which are almost as high, are on other sides of the lake and may be reached on foot.

111

BRITISH COLUMBIA AND THE PRAIRIE PROVINCES

nipeg is by far the largest city, but Brandon and St. Boniface are growing.

The soil of Manitoba is very rich, yielding wheat, oats, rye and barley abundantly. In the north are great forests of white spruce and jack pine, and from the north too comes a rich harvest of furs, and a promising return of minerals from the famous Flin Flon copper ore and the rich Mandy copper claim in The Pas. Although the winter climate is severe yet there is a great amount of bright sunshine and but little cloudy weather. Rain falls in June and July.

The Climate of the Province

British Columbia is in some respects the most favored part of Canada. Within its boundaries are reproduced all the varied climates of the Dominion and almost every natural feature, while some of its climatic and geographical conditions are peculiar to the province. Extending from the Rockies to the Pacific and from the 49th to the 60th parallel of latitude, it has an area of 366,255 square miles, about three times the size of Italy. The many islands of the Pacific Coast, notably Vancouver Island (area 12,408 square miles) and the Queen Charlotte group, are included in the province. They are noted for their temperate climate and abundant natural resources. The mines, timber areas, fisheries and agricultural resources of the province are remarkable for their quality and extent.

Mountains and Valleys

British Columbia is essentially a mountainous country, comprising practically the entire width of the Cordilleran belt of North America. The chief system in this belt are the Rocky Mountains proper on the northeast side, and the Coast Range on the southwest or Pacific side. Between these are lower ranges running southeast and northwest. Vancouver Island and the Queen Charlotte group are remnants of still another range now almost entirely submerged in the Pacific. The highest peak in the Canadian Rockies is Mount Robson; Mount Fairweather on the International Boundary is the highest peak in the province. Other high peaks include Columbia, Forbes, Assiniboine, Bryce. Passes over the Rockies are many: the South Kootenay, Crow's Nest, Kicking Horse (traversed by the main line of the Canadian Pacific), the Yellow Head Pass (used by the Canadian National) and the Peace River Pass. The Coast Range renders the coastline of British Columbia remarkable not only for its extent (7,000 miles) caused by deep fjord-like indentations, but also for its great beauty as the mountains rise from the water's edge to a height of 5,000 to 8,000 feet.

Mountains imply valleys and it is in these valleys that the agricultural wealth of the province of British Columbia is produced. The Okanagan Valley, stretching for eighty miles north and south, is famed for its apples, cherries, apricots and peaches. The Fraser Valley floored with rich alluvial soil brought down by the Fraser River grows immense crops of hay and grain and supports a large dairy industry. The benchland on its borders is well adapted for the growing of berries and other small fruits, and here the Japanese have entrenched themselves very strongly.

Great Mineral Wealth

Mountains often contain minerals, and the settlement on the mainland of British Columbia was due largely to a "gold rush" of the early sixties up the Fraser Canyon to the Cariboo country. On Howe Sound rich copper is worked by the Britannia Company; at Stewart on the Alaskan border is the Premier mine, one of the richest small mines known, producing both gold and silver; at Anyox near Prince Rupert more copper is to be found; near the southeastern corner of the province is the greatest zinc-lead mine in the world, the Sullivan mine, and not many miles away at Trail is a great smelter where very pure zinc and lead are made from Sullivan ore.

British Columbia has the largest area of salable timber of any country. Three-quarters of her area is covered with valuable timber and the forests include yellow pine, Douglas fir, red cedar, hem-

PHOTOS, IMPERIAL OIL LIMITED

PROSPECTING with Geiger counters in the uranium-rich area of northern Saskatchewan.

STAKING A CLAIM near Uranium City, center of the Saskatchewan development.

ALUMINUM COMPANY OF CANADA, LTD.

POWER-HOUSE SITE at Kemano, British Columbia. It is part of the vast Kitimat project under which water-power resources are being developed to supply the aluminum industry.

lock, balsam and spruce. Most majestic of all the western forest trees is the Douglas fir, which grows at times to a height of nearly two hundred feet, has a girth exceeding thirty feet and a finished lumber which almost equals oak in beauty.

A large part of the commerce of British Columbia is derived from the sea. The chief product is salmon caught along the coast and in the rivers and inlets. Large canneries are in operation employing many fishermen, white, Indian and Japanese. The headquarters of the halibut fishery are at Prince Rupert and from this point hundreds of boats set out.

Early in the nineteenth century four nations, Spain, Russia, Great Britain and the United States, claimed the "Oregon Country." Spain surrendered her claims to any land north of the present California to the United States and Russia withdrew within the present Alaska, leaving only two claimants for the vast region. Both based their claims chiefly upon exploration, Captain Cook (1778) and Captain George Vancouver (1792–94) had explored the coast. Alexander Mackenzie, Simon Fraser and David Thompson, all of the North-West Fur Company, had reached the Pacific overland. On the other hand, Captain Robert Gray of Boston had entered and named the Columbia River in 1792, and Lewis and Clark had floated down the Columbia. In addition the United States had succeeded to whatever claim Spain had had. Both nations established fur-trading posts, and finally, in 1846, the region was divided by prolonging the 49th parallel, leaving, however, all of Vancouver Island to Great Britain. The island's southern tip is below the parallel.

Vancouver Island was proclaimed a British colony in 1849, and, in 1858, following a gold rush, the territory on the mainland was proclaimed as British Columbia. Eight years later the two districts were joined in administration. In

SASKATCHEWAN GOVERNMENT

THE DAY'S WORK IS DONE and both cowboys and horses turn happily homeward on a Saskatchewan ranch. The weathered log bunkhouse, in the grove of trees, is still in use.

NATIONAL FILM BOARD

EXPERT INSPECTION OF PELTS BEFORE A FUR AUCTION BEGINS

Buyers look closely at muskrat skins at an auction room in Winnipeg. They are permitted to do so until the sale begins. Only registered individuals and representatives of firms known to be members of the fur industry may offer bids at a raw fur auction. Auctions, which speed up the trade in pelts, date back to the early days of the Hudson's Bay Company.

NATIONAL FILM BOARD
LUMBERJACKS stand on springboards wedged in the trunk to help them cut a spruce.

STANDARD OIL CO. (N. J.)
AN A-FRAME DERRICK lowers two hundred tons of logs into the booming ground.

NATIONAL FILM BOARD
CANADIAN LUMBER lies on the docks of a port in British Columbia before being shipped all over the world. The lumber industry provides Canada with one of her largest exports.

BRITISH COLUMBIA AND THE PRAIRIE PROVINCES

1871 British Columbia entered Confederation on condition that the government would bring a railway through the mountains, and also introduced responsible government. At that time her population was only 36,000, but with the coming of the railway the increase was rapid. In 1951, according to the census that year, it was 1,165,210. The percentage of British born (over thirty) is larger in this province than in any other.

The development of her rich natural resources has led British Columbia into a difficulty. Facing eastward, she looked to the Orient to supply her labor. By 1884 nearly 10,000 Chinese were in the Province working cheaply and sending their savings back to China. In 1902 the head tax on Chinese was raised to $500, reducing immigration from China. However, people from Japan and India continued to come and make their homes. Many of them went into such occupations as gardening, lumbering and fish-canning.

The chief cities of the province are Victoria, the capital, on Vancouver Island, Vancouver on the mainland, New Westminster on the Fraser and Nanaimo in Vancouver Island. At Point Grey the growing University of British Columbia occupies a magnificent site, which was chosen for it after much discussion.

NATIONAL FILM BOARD

AN AUTOMATIC DRILL makes blasting holes for dynamite, in the great open-pit mines at Flin Flon, Manitoba. The ore is rich in such metals as zinc, copper, gold and silver.

FEDERAL GOVERNMENT GRADERS in the Winnipeg stockyards stamp beef carcasses according to structure, age and weight. About 500,000 cattle a year go through the stockyards.

SALMON NETS must be hauled in with care so that they will neither break nor tangle. British Columbia's salmon fleet, the largest in the world, consists mostly of small vessels.

A DAY'S CATCH of salmon is emptied onto a bed of ice in a cannery barge. Whatever parts of the fish are left over from canning will be used in other products, such as fertilizers.

BRITISH COLUMBIA AND THE PRAIRIE PROVINCES

BRITISH COLUMBIA AND THE PRAIRIE PROVINCES: FACTS AND FIGURES

This chapter includes the three prairie provinces of Manitoba, Saskatchewan and Alberta and the coastal province of British Columbia.

BRITISH COLUMBIA has a total area of 366,255 square miles; land area, 359,279 square miles, including Vancouver Island (12,408 square miles). The total population is 1,165,210. A bill to annex the Yukon Territory to British Columbia has been considered. Administered by a lieutenant governor and a ministry responsible to a Legislative Assembly of 48 members. Representation in Canadian Senate, 6; in House of Commons, 22. Manufacturing, forestry, mining and agriculture are important occupations. Forests include fir, cedar, hemlock, spruce and pine. Leading minerals are lead, zinc, gold, coal, copper and uranium; province ranks third in value

COURTESY CANADIAN NATIONAL RAILWAYS

POCAHONTAS POST OFFICE, ALBERTA

In rural communities and in new sections the post office is the center of community life.

of mineral production. There are 3,526 manufacturing establishments; leading products—lumber, pulp and paper, fish, ships and ship repairs, fertilizers, plywood, beverages, fruits and vegetables. Population about four-fifths Protestant. Provincial public education is compulsory for ages 7 to 15; 164,212 pupils in public schools; 6,730 students at the University of British Columbia and Victoria College. Railways, 4,813 miles; surfaced roads, 10,310 miles; 275,000 telephones; 186,108 radio receiving sets. Population of chief cities: Vancouver, metropolitan area, 530,728 (city proper, 344,833); Victoria (capital), metropolitan area, 104,303 (city proper, 51,331); New Westminster, 28,639; North Vancouver, 15,687.

MANITOBA has a total area of 246,512 square miles; land area, 219,723 square miles. Population is 776,541. Province is administered by a lieutenant governor appointed by the Federal Government and a ministry responsible to a Legislative Assembly of 57 members. Representation in Canadian Senate, 6; in House of Commons, 14. Agriculture and stock-raising important; chief farm products are wheat, oats, barley and milk. The leading minerals are copper, zinc, cement and gold. Fresh water fisheries are extensive. There are 1,600 manufacturing establishments, devoted chiefly to slaughtering and meat-packing, railway equipment repairs, butter- and cheese-making, printing and brewing. Railways, 5,344 miles; all-weather roads, 9,109 miles; 201,000 telephones in use. 128,868 pupils in public schools; in addition to provincial university and 5 affiliated colleges there are several private colleges. Population of chief cities: Winnipeg (capital), metropolitan area, 354,069 (city proper, 235,710); St. Boniface, 26,342; Brandon, 20,598.

SASKATCHEWAN has a total area of 251,700 square miles; land area, 237,975 square miles. Population of province is 831,728. Administered by a lieutenant governor appointed by the Federal Government and a ministry responsible to a Legislative Assembly of 52 members. Representation in Canadian Senate, 6; in House of Commons, 17. Agriculture, manufacturing and mining are the chief industries. Leads Canada in the production of wheat, oats and rye. Copper, zinc, gold, coal and sodium sulfate are the chief minerals. There are 1,001 manufacturing establishments producing chiefly meats, flour and feed, beverages, butter and cheese and printed material. Railways, 7,011 miles; surfaced roads, 13,207 miles; 122,987 telephones. 172,738 pupils in public schools; also the University of Saskatchewan and 2 normal schools. Population of chief cities: Regina (capital), 71,319; Saskatoon, 53,268; Moose Jaw, 24,355; Prince Albert, 17,149.

ALBERTA has a total area of 255,285 square miles; land area, 248,800 square miles. Population is given as 939,501. Administered by a lieutenant governor appointed by the Federal Government and a ministry responsible to the Legislative Assembly of 57 members. Representation in Canadian Senate, 6; in House of Commons, 17. Agriculture, manufacturing and mining are chief industries. Chief farm products are wheat, milk, barley and oats. Coal, crude oil (95% of the country's production) and natural gas are the chief mineral products. There are 1,685 manufacturing establishments producing chiefly meats, beverages, flour and feed and lumber. Railways, 5,805 miles; surfaced roads, 18,066 miles; Edmonton is terminal for air flights to the north and across the Pacific; telephone-wire mileage, 341,781; 113,000 telephones. 192,227 pupils in public schools; 7,871 correspondence-school students; in addition to provincial university and technological institute there are private colleges and normal schools. Population of chief cities: Edmonton (capital), 159,631; Calgary, 129,060; Lethbridge, 22,947; Medicine Hat, 16,364.

CANADIAN CITIES

Great Variety and Striking Contrasts

Even when Canada was chiefly an agricultural country, it had great and small cities and prosperous towns. During recent years it has also become a leading industrial country, exporting manufactured goods as well as farm products to all parts of the world. This has affected the growth and character of its older cities and led to the establishment of new ones. Yet in many places the atmosphere of days gone by still lingers, amid all the bustle of industrial expansion. From St. John's, Newfoundland, to Victoria, over three thousand miles away, most of the principal cities are strung at irregular intervals almost upon a line. Each of them has something distinctive and interesting to offer the visitor.

IT seems fitting to begin an east-to-west tour of Canada's chief cities by visiting one whose history goes back further than the earliest days of British colonization. When North America was still largely unknown to white men, and Henry VIII sat on the throne of England, St. John's, Newfoundland, was already a fishing center with a thriving harbor. By the time Newfoundland became a British colony, in 1583, St. John's was well established. In the course of time it became known far and wide as the commercial and processing headquarters for the great Atlantic cod fisheries and as a major shipping point. In 1949, when Newfoundland joined Canada as the tenth province, St. John's closed its career as the capital of Britain's first overseas colony and became the capital of the new province.

Among the city's notable buildings is the Anglican Cathedral, a fine Gothic structure designed by Sir Gilbert Scott, a famous nineteenth-century English architect. Another is the Roman Catholic Cathedral, which stands in a commanding position on a hill. Government House is a replica of Admiralty House in Plymouth, England. Memorial University College and the Newfoundland Hotel are more modern buildings.

The way from the ocean to St. John's harbor passes between beetling cliffs 500 feet high, known as the Narrows. From Cabot Tower, on the north side of the Narrows, Marconi received the first transatlantic wireless message in 1901. Eighteen years later, St. John's was the starting point of another great exploit—the first nonstop flight across the Atlantic, made by Alcock and Brown.

A journey of almost 500 miles brings us to the mainland city of Halifax, founded in 1749. It is the capital of Nova Scotia, Atlantic headquarters of the Royal Canadian Navy and eastern Canada's chief winter port. From the old stone citadel on a dome-shaped hill there is a superb view of the city and the magnificent harbor. The city is protected by powerful forts and batteries, for Halifax has been an important naval and military base since a royal dockyard was established there in the great days of sail, two hundred years ago. One of the largest bridges in the British Commonwealth is now being built to span the harbor and connect the city with the town of Dartmouth.

The city's two cathedrals, the new public library, Dalhousie University and many other fine buildings help to give Halifax dignity and charm. Halifax also boasts the oldest Protestant church building in Canada, St. Paul's Church. The public gardens and parks are delightful, and an arm of the sea which reaches behind the city, known as North-West Arm, is one of the finest aquatic playgrounds to be found anywhere.

Prince Edward Island's only city is a small and attractive one, Charlottetown, founded as Port La Joie before 1750. Its picturesque old Colonial Building was the scene of the historic meeting in 1864 which led to the confederation of Canada three years later. Charlottetown has lobster and oyster fisheries; and with several delightful seaside resorts only a short distance away, it attracts a great many summer tourists.

Although Fredericton is the capital of New Brunswick, Saint John is the largest city in the province. It is also the oldest incorporated city in Canada, with a Royal Charter dated 1785. Originally built as a fortress on solid rock at the mouth of the Saint John River, it was the landing place of ten thousand United Empire Loyalists who left the United States during the War of Independence in order to remain under British rule. Because of serious fires which swept the city in recent times, Saint John today consists mostly of modern buildings. Its harbor is notable not only for its varied trade but also for the Reversing Falls, which produce surging torrents at each change of the tide.

The motto of the province of Quebec is *Je me souviens,* "I remember." That is what the ancient stones of Quebec City seem to murmur as one explores the narrow, crooked streets and the old buildings of the lower town, huddled at the base of the great rock upon which the city stands. To reach the upper town, one can climb flights of steps or winding roads cut out of the rock, or ride in huge elevators. Here in the upper town are Laval University and the Grand Seminary, the Château Frontenac, and Dufferin Terrace, a famous boardwalk set out from the side of the cliff. Here are the Wolfe-Montcalm monument and the monument to Champlain, who established a trading post at Quebec in 1608, making it the first permanent settlement in Canada. And dominating all, high above the river, crowning what is surely the most dramatic city site on the continent, is the Citadel. From the battlements one looks below to the roofs of the lower town and across the broad and busy St. Lawrence to the Isle of Orleans and the city of Lévis on the farther shore. Beyond are range upon range of mountains, the summits rounded

THE PROTECTED HARBOR of St. John's is a haven for the Newfoundland fishing fleet as well as for other kinds of vessels. Around the harbor the city is spread out on hills.

NATIONAL FILM BOARD

THE TOWN CLOCK crowns Citadel Hill, the highest point in Halifax. From this vantage point there is a sweeping view of the city's business section and the sheltered bay.

by time, fading into the blue distance.

Quebec City is such a fascinating place in which to wander, so full of the echoes of times past, that the casual visitor is likely to overlook its day-to-day activity as a provincial capital, as a religious and educational center, as the home of many industries and as a port. But it is all of these things as well as a tourist resort, and its people live amid a mixture of past and present.

The same is largely true of Canada's biggest city, Montreal, which lies 180 miles farther into the heart of the continent along the St. Lawrence. Called Hochelaga by the Indians and Ville-Marie by the early French settlers, it takes its present name from Mont Réal (Mount Royal), a volcanic hill rising about 800 feet from the island of Montreal. The city is built on a series of terraces on the hill's lower slopes. Above the harbor comes the business district, with the Bank of Montreal as the most stately of many impressive buildings. Nearby, in the Place d'Armes, is Hébert's striking

EDWARD CORNWALLIS, English soldier, still looks over Halifax, which he founded.

MARKET SLIP at Saint John, New Brunswick, where a variety of supplies reach the city. It is a major Canadian port, ice-free the year round, with one of the world's largest dry docks.

bronze figure of Maisonneuve, founder of the city. The principal stores and hotels are halfway up the slope; the main residential districts are higher still. The topmost part of the mountain has been left more or less in its natural state. It forms a public park commanding lovely views.

Four centuries ago Jacques Cartier planted a crude wooden cross on the summit of Mont Réal. Today the chief landmark by night is a great illuminated cross at the same spot. No symbol could be more fitting, for Montreal is a city of religions. Nearly all beliefs have a place there. However, scores of buildings—ancient shrines, convents, churches, hospitals, seminaries—show that here is a bastion of the Roman Catholic Church. Outstanding among the religious buildings are St. James Cathedral, modeled on St. Peter's in Rome; the parish church of Notre Dame, which seats ten thousand; the Anglican Cathedral, Christ Church, St. James United Church and the quaint Bonsecours Church. The hospital of the Hôtel Dieu has existed for over three hundred years, but not in its present building. The city's most interesting older buildings are perhaps the Seminary of St. Sulpice and the Château de Ramezay, once the home of the French governors and now a museum. McGill University and the University of Montreal are centers of higher education and research.

Montreal has grown from a settlement relying mainly on the fur trade into an industrial, commercial and financial metropolis of over a million people. It serves a huge area and has world-wide connections. It is the headquarters of Canada's two transcontinental railway systems. Many banks, industrial corporations and other national enterprises have their head offices there. However, it is shipping in all its branches that is the most important industry. Montreal's unique position as a sheltered seaport, a thousand miles from the open sea, makes it the chief link between the world's ocean trade, the vast water-borne trade of the Great Lakes, and the wealth of the West. When the St. Lawrence Seaway is completed, the harbor and docks of Montreal will probably be even busier than they are today.

Verdun, Outremont, Westmount, Laprairie and Lachine, neighboring cities and towns forming the suburbs of Montreal, are delightful, with many attractive homes and gardens. As in the city itself, their people of different national origins and creeds mingle in harmony and work together for the common good.

About a hundred miles west of Montreal, just inside the province of Ontario where it borders Quebec Province, is Ottawa, the nation's capital. It occupies a magnificent site on the Ottawa River, chief tributary of the St. Lawrence. Once known as Bytown, after Colonel John By of the Royal Engineers who built the Rideau Canal in the 1820's, this former lumbering center became the city of Ottawa in 1854. Three years later, much to the dismay of the hopeful rival cities of Montreal, Quebec and Toronto, Queen Victoria selected it as the capital. Although Ottawa is first and foremost a seat of government and a city of imposing official buildings, it is also a place of rivers and rocks, of bridges and locks and canals, streams and cascades. The Parliament Buildings stand on a high bluff above the Ottawa River, with the distant ranges of the Laurentian Mountains as a backdrop for their graceful towers and buttresses. Almost as familiar to visiting statesmen and diplomats are the towers of the nearby Château Laurier Hotel, a gem of period architecture. Rideau Hall, the governor general's residence, is a large, rambling building surrounded by attractive grounds.

Near the center of the city are the seething Chaudière Rapids; elsewhere the Gatineau River and the Rideau Canal wind past sections of well-kept lawns and multihued flower beds. Ottawa is perhaps at its best on a fine May morning.

THE PROVINCE BUILDING in Charlottetown, home of the Prince Edward Island legislature. When the present city was laid out in 1768, it was named Charlotte(town) for George III's wife.

BOTH PHOTOS, CANADIAN NATIONAL RAILWAYS

THE CHATEAU FRONTENAC, one of the landmarks of the upper town of Quebec City. Dufferin Terrace, the famous railed boardwalk, runs along the cliff top beside the hotel.

Then the tulips which are an annual gift from Queen Juliana of the Netherlands are in full bloom around Ottawa's buildings and along its waterways; the sun glints on the swirling river and its rocky banks, and the city's elm-lined streets are vistas of tender green.

Connected with Ottawa by the Rideau Canal is the old city of Kingston. At first it was a French fort; later it became a British one. Fort Henry, built during the War of 1812 to protect the naval dockyard which had recently been established there, is now a national historic site. The city's many fine old buildings include the City Hall, St. George's Cathedral and the Royal Military College, which corresponds to Sandhurst in England and West Point in the United States. Queen's University is noted for the quality of its teaching and the harmonious beauty of its gray limestone buildings.

A "Forest" City

On his first visit to Toronto, a well-known British diplomat looked out of his window high up in a big hotel and said, "Why, it's not a city at all; it's a million people living in a forest!" All kinds of trees are to be found in Toronto's downtown streets and in its many parks and beautiful residential sections and suburbs, such as Rosedale, Forest Hill Village, Alexandra Wood and the Kingsway. Canada's second largest city is growing at such a pace that some of its many trees have made way for expressways, factories, office buildings and blocks of apartments. However, those that remain are a lovely natural asset.

Toronto is the largest unit in a confederation of thirteen neighboring municipalities. It is the political and financial capital of Ontario, a leading industrial and commercial center and a lake port with a good harbor and several miles of docks and wharves on Lake Ontario. Its citizens can name a long list of outstanding local features that include the only subway in Canada; the largest university, the largest hotel (the Royal York), and the highest building (the Canadian Bank of Commerce) in the British Commonwealth; the busiest mining-stock exchange in the world; the largest annual fair in the world (the Canadian National Exhibition); and more telephones per person than any other city of the Commonwealth.

PHILIP GENDREAU

CHAMPLAIN MONUMENT in Quebec, the city founded by the French explorer-soldier.

Toronto is a sprawling, bustling, go-

OLD-FASHIONED VICTORIAS carry sight-seers to the park at the top of Mount Royal. The bustling metropolis of Montreal is spread out below, girdled by the St. Lawrence River.

ST. JOSEPH'S ORATORY, Montreal, founded in 1904 by a humble lay brother, André.

getting business city. But at the same time it is a cultural center of note. It prides itself on its fine museums, art galleries and libraries, supports its own symphony orchestra, and appreciates the opera, the ballet and the theater. It also is proud of its historical heritage. Old Fort York, built by Governor Simcoe in 1794 and recently restored, is an interesting relic of colonial days. Another historic spot is the homestead of William Lyon Mackenzie, first mayor of the city and leader of the Upper Canada Rebellion of 1837. However, the leading tourist attraction is a storybook "castle" named Casa Loma. This enormous, fantastic structure was built as a residence by the late Sir Henry Pellatt at a cost of nearly $3,000,000. It is now operated as a showplace and ballroom for charitable purposes.

Hamilton, at the western end of Lake Ontario, is sometimes called "the Pittsburgh of Canada" because it is the coun-

IN DOWNTOWN MONTREAL the Laurentien Hotel displays one of Canada's great products, aluminum. The walls are faced with more than 70,000 square feet of the silvery metal.

try's foremost center of the iron and steel industry. By night the chimneys and blast furnaces light up the surrounding sky. Hamilton is also an attractive place by day. It has good public buildings and residential districts, and its celebrated open-air market handles much of the luscious output of the Niagara fruit belt. McMaster University, on the city's outskirts, enjoys a reputation that is all out of proportion to its small size. The exquisite rock gardens on Hamilton "Mountain" attract visitors from far and wide.

Another city of southwestern Ontario with lovely surroundings is London, an old pioneer settlement. Governor Simcoe wanted to make it the capital of Upper Canada in 1792, but he was overruled in favor of Toronto. The locomotive on the civic coat of arms is no mere figurehead, for modern London is an important railway center. This delightful and compact city is big enough to be interesting in most of the ways that matter, but it is

OUTSIDE STAIRCASES along an old Montreal street. It is illegal to erect them today.

PARLIAMENTARY LIBRARY in Ottawa. It is a circular structure in the Gothic style.

small enough to have a particular charm of its own. Its natural beauties are numerous. It would be difficult, for instance, to find a more beautiful campus than that surrounding the University of Western Ontario.

Like Hamilton, the city of Windsor is a hive of industry dominated by huge plants and shops. This great automobile-manufacturing center lies along the Detroit River opposite Detroit, Michigan, to which it is connected by bridges, tunnels and ferries. Windsor has an assured place in history, for here is the Bâby House which was used as a headquarters by General William Hull when he invaded Canada from the United States in the War of 1812. It is said to be the oldest brick house in Upper Canada. Before the American Civil War, Windsor spelled freedom for many of the slaves using the "underground" escape routes from the South to Canada.

Ontario has more cities than any other Canadian province, and all are interesting in one way or another—Brantford and Brockville; Guelph and Galt; Peterborough with its tremendous lift lock on the

BOTH PHOTOS, NATIONAL FILM BOARD

SUPREME COURT BUILDING in Ottawa. Like most of the other federal government buildings, this one has a magnificent location on the bluffs above the Ottawa River.

LADY GREY DRIVE in Ottawa guides visitors along a wandering route following the Ottawa River and past the Royal Canadian Mint (right) where the country's coins are made.

Trent Canal; St. Catherine's and its mineral springs; Stratford, famed for its annual Shakespeare festival; Sarnia, city of oil refineries; Sudbury and Timmins, the mining cities of the north; Kitchener, once called Berlin; the twin lake-head cities of Fort William and Port Arthur, marked by rows of enormous elevators crammed with grain from the west awaiting shipment overseas. These twin cities form a vital link between eastern and western Canada. In western Canada our first important stop is Winnipeg, capital of Manitoba. This largest city of the prairies and the fourth largest in Canada is at the junction of the Assiniboine River and the Red River of the North, in the center of Canada's richest farmlands. Winnipeg began as a small Hudson's Bay Company post called Fort Garry, about 1820. Its early days were marked by the bitter struggle between rival companies for control of the rich fur trade. The city grew very slowly; in 1871 the population was under 250. Then the Canadian Pacific Railway arrived in 1881, followed by a rush of settlers, and the population began to soar.

Into the railway yards of Winnipeg come great trainloads of wheat, oats, barley, rye and flax; the turbulent but orderly Grain Exchange buys and sells them. Early settlers planned the city on a generous scale. Its wide, airy streets are in agreeable contrast to the congested streets of most other great cities. The buildings, both public and private, are admirable. Outstanding are the beautiful Legislative Building, opened in 1920, and the Hudson's Bay Company's department store, which occupies a whole block. The company has landscaped and presented to the city the ivy-covered tower of the old fort, last reminder of earlier, stormier days.

Though founded in 1882, Regina, the capital of Saskatchewan, is thoroughly modern in appearance, for the original

LAKE FRONT, TORONTO. Docks and warehouses stretch along the water's edge. On the skyline are the Royal York Hotel (left rear) and (right of hotel) the tower of the Bank of Commerce.

SEMICIRCULAR SECTIONS with Ionic columns emphasize the Greek style of Convocation Hall, Toronto University. The institution has an especially beautiful and spacious campus.

A RECONSTRUCTION OF FORT YORK on the old site in present-day Toronto. The French built the fort in 1749 and the British captured and destroyed it ten years later.

A GREAT LAKES FREIGHTER takes on grain from the huge elevators at Fort William, Ontario. On Thunder Bay on the northwest shore of Lake Superior, the city has a fine natural harbor.

BLAST FURNACES in busy steel mills make the night sky glow over Hamilton, Ontario. The attractive city at the head of Lake Ontario is the center of Canada's iron and steel industry.

city has disappeared. On a June day in 1912, it was laid in ruins by the most devastating tornado ever to hit Canada. After the storm, ungainly temporary buildings were put up. Gradually these were replaced, until the city became a bright, thriving place with excellent buildings, large park areas and a profusion of trees. This pleasant city is the distributing and marketing center for a huge prairie region and is the home of several flourishing industries, such as woodworking, oil refining and printing.

Northwest of Regina, in the heart of the hard-wheat belt, is Saskatchewan's second largest city, Saskatoon. It also is a roomy, modern city which has grown steadily from small beginnings into a prominent trading center for farm products. There are extensive flour mills and a few other manufacturing plants. Spread along both banks of the South Saskatchewan River, the city is linked together by several large bridges. Near the river are tree-shaded parks and the attractive campus of the University of Saskatchewan.

The main line of each of Canada's great transcontinental railway systems serves one of the two major cities of Alberta. From Regina, the Canadian Pacific Railway goes by way of Moose Jaw and Medicine Hat to Calgary. Farther north, the Canadian National Railway connects Saskatoon with Edmonton, the capital of Alberta. Like many other Canadian cities, Edmonton's early fame developed from the fur trade. The city takes its name from Fort Edmonton, a Hudson's Bay Company trading post set up in 1795 on the North Saskatchewan River. The city has always been prosperous because it lies near good farm country and deposits of coal, gold and pitchblende. Its extraordinary growth lately is due to the discovery and development of oil fields in Alberta. Chemical production is a leading industry, and the presence of great quantities of natural gas is another industrial asset.

For good reasons, Edmonton has been called "the gateway to the north" and "the crossroads of the world" because it is a transportation hub of the first importance. No less than thirteen lines of railway enter the city. Here, too, is the chief junction for the air routes to the great tracts of country being opened up in the north.

On the western edge of the Alberta prairies, not far from the Rocky Mountains, the foothill city of Calgary is the commercial center of the oil boom. Calgary was first settled as a Royal North West Mounted Police post in 1875. It soon became the local capital of the ranching country surrounding it and carried on a huge trade in cattle, sheep and horses. Standing 3,500 feet above sea level, where the Bow and Elbow rivers meet, the city is only 80 miles east of the Rockies. The westerly view at sunset is almost unbelievably beautiful.

Calgary is rich in parks and public gardens. In Dinosaur Park can be seen life-sized models of the fearsome reptiles that roamed Alberta millions of years ago. The domain of the ancient dinosaurs is today the home of the wild horses which snort and plummet into the ring at the annual Calgary Stampede. This world-famous show is a colorful mixture of agricultural fair and thrilling competition. There are parades of cowboys and Indians, chuck-wagon races, and steer-roping and riding contests. Most exciting of all are the competitions between three-man teams attempting to rope, halter and ride unbroken horses.

Smaller Alberta cities on the route west are the industrial center of Medicine Hat and Lethbridge. The latter is an impor-

THE SHAKESPEAREAN FESTIVAL held every year has made Stratford, Ontario, a Mecca for lovers of the theater from all over North America. Performances are of high professional caliber.

tant air-line junction which took its name from William Lethbridge. He was the head of a company operating local coal mines and a narrow-gauge railway nicknamed "The Turkey Trail."

Few cities in the world can boast a natural setting to equal that of Vancouver, British Columbia, Canada's third largest city and her greatest seaport. Vancouver lies on the Gulf of Georgia, just north of the Canada-United States border. The lofty mountains of the Coast Range rise behind the city and from practically every point there are lovely vistas of sea and mountain. Originally a small mill town named Granville, Vancouver took its present name in 1886. That same year it was wiped out by fire. It was speedily rebuilt and has been growing at a great pace ever since. Today it is the western terminal of the Canadian Pacific and Canadian National railways. Its great harbor handles a higher tonnage of shipping than any other port in Canada.

STANDARD OIL CO. (N. J.)

LIFELINES OF INDUSTRY flow through Winnipeg, Manitoba, one of the largest railroad centers on the continent. There are 270 miles of track in the Canadian Pacific freight yards.

A TIME-WORN GATE—the remains of old Fort Garry. There Hudson's Bay Company traders once met. Today the spot is a national historic site, standing in the midst of downtown Winnipeg.

BROADWAY BRIDGE, one of several spans that link Saskatoon, Saskatchewan. Spread out along both banks of the South Saskatchewan River, the city is spacious and modern.

SASKATCHEWAN'S BEAUTIFUL CAPITAL. Regina is on Waskana Creek, which here forms a pretty lake. The large building with a cupola houses the legislature of the province.

AN ATTRACTIVE ROW of comfortable modern homes in Lethbridge, Alberta. The city is well planned. Streets are wide and lined with trees; and householders take pride in their gardens.

DOWNTOWN IN BUSY EDMONTON. In the central part of Alberta, at the southern end of the Alaska Highway, the city serves an area rich in mineral resources—coal, oil, gas.

SAWDERS FROM CUSHING

DINOSAUR IN CONCRETE in the Fossil Gardens of Prince George Island Park in Calgary, Alberta. Remains of such beasts have been found not far away, in the Red River valley.

CANADIAN PACIFIC RAILWAY

CALGARY, the city where the Bow and Elbow rivers meet. Long famous as the center of a stock-raising area, its prosperity has been rurther increased by the rich oil fields near by.

THE CITY HALL of Vancouver, British Columbia, is a spacious structure in modern setback style, with a clock on each side of the top. Shrubs are massed near the entrance.

Immense cargoes of wheat, lumber and other commodities are shipped to the Orient, to the United States and South America, and to Europe by way of the Panama Canal. Besides being a major center of trade, industry and transportation, Vancouver is one of the chief pleasure resorts and residential cities of the northwest Pacific area, with numerous beaches and delightful suburbs. The climate is mild and the grass stays green all the year round.

Thousand-acre Stanley Park is the best known of Vancouver's many fine parks and open spaces. Perhaps the most striking building in the city is the Hotel Vancouver, opened in 1939 and now a leading social and business center. The city's Chinese quarter houses many thousands of Chinese people and in North America is second only to the Chinese quarter of San Francisco. A splendid feature that arrests the eye of a visitor is the Lions Gate Bridge over the harbor mouth. Over 1,500 feet long, it is the longest and largest suspension span in the British Commonwealth. The University of British Columbia occupies a beautiful site on Vancouver's outskirts.

Twelve miles from Vancouver is the city and port of New Westminster, a distributing center for the fruit of the Fraser Valley. It was founded in 1859 by the Royal Engineers and was for a time the capital of British Columbia. Fish processing is the main industry at Prince Rupert on Kaien Island, five hundred miles northwest of Vancouver. Each year millions of pounds of fish are processed in its colossal plants.

When Vancouver Island became a crown colony, Victoria was selected as its capital. It remained the capital when the island united with the mainland colony of British Columbia. Today it is a beautiful, quiet, well-kept city with few echoes of its boisterous past. However, at the time of the Cariboo gold rush, about a century ago, it was a lusty tent town of ten thousand people, most of them gold-crazy and many of them violent. This hardly sounds like the place of which

BASKING ON THE BEACH at Stanley Park, in Vancouver. The well-tended park covers nine hundred acres. Besides the wide beach there are a zoo and extensive gardens.

CANADIAN NATIONAL RAILWAYS

VICTORIA & ISLAND PUBLICITY BUREAU

THE GRACIOUS CAPITAL of British Columbia—Victoria—viewed from the top of the Parliament Building. Its many parks—Beacon Hill, Gorge—are beautifully landscaped.

Rudyard Kipling said, "I tried honestly to render something of the color, the gaiety, and the graciousness of the town and the Island, but only found myself piling up unbelievable adjectives."

Victoria—with its flowers and gardens and magnificent views, its Empress Hotel and attractive Parliament Building set amid velvety lawns, its Scottish tweeds to wear and its English crumpets for tea—is sometimes called "a bit of old England in new Canada." There is some truth in this, but not very much. Victoria is essentially a Pacific coast city with its interests and connections centered in the Pacific coast. It is important not only as one of the most charming residential cities in the world but also as the political capital of its province and one of Canada's busiest ports.

The census of 1951 showed that Canada had 55 cities with a population of 20,000 or more. This does not include many suburban areas around the bigger cities, which have large populations but are not themselves incorporated as cities. One city, Montreal, had over 1,000,000 people. Toronto had over 600,000, Vancouver over 300,000. There were three cities with more than 200,000 inhabitants—Hamilton, Ottawa and Winnipeg. Four others had more than 100,000—Calgary, Edmonton, Quebec and Windsor. Halifax, London, Regina, Saint John (New Brunswick), St. John's (Newfoundland), Saskatoon, Sherbrooke, Verdun and Victoria claimed between 50,000 and 100,000 people. The populations of twenty-four cities ranged from 25,000 to 50,000. Twelve cities had from 20,000 to 25,000 persons. About a third of the people of Canada live in the larger cities. This proportion is likely to increase because the cities have a strong attraction for both the Canadians and the immigrants who are entering the country from Europe in large numbers.

BY R. D. HILTON SMITH

WHITEHORSE—ON LEWES RIVER—THE YUKON'S LARGEST TOWN
Air service, the navigable Lewes River, the Alaska Highway, a railroad to the port of Skagway, Alaska, and the radio keep remote Whitehorse in touch with the rest of the world.

PHOTOS, NATIONAL FILM BOARD

MAIN STREET—A CORD OF WOOD, FAMILIAR SIGNS, A CHURCH
People for miles around and countless travelers depend on the conveniences of Whitehorse's Main Street. Miners, trappers, hunters from all directions come into town to buy supplies.

THE ROOF OF NORTH AMERICA
The Yukon and the Northwest Territories

Stretching from the sixtieth parallel all the way to the North Pole, the Yukon and the Northwest Territories make up about two-fifths of the total area of Canada. Many sections to the far north are still unknown, though white men saw the eastern shores not long after Columbus discovered the New World. It is the eastern regions, however, that remain aloof in arctic cold. The western sections have proved to be more hospitable to man and, since just before the opening of the twentieth century, have revealed a wealth of mineral treasure. Here are a number of mines, oil wells and settlements.

FOR more than three centuries the history of the territories was a story of gallant adventure. Soon after North America was discovered, bold mariners began seeking a route—the elusive Northwest Passage—through the unknown icy seas at the "top" of the continent. Between 1576 and 1587 English explorers discovered Davis Strait to the west of Greenland, and in 1610 Henry Hudson entered the bay named for him. For twenty years thereafter a series of expeditions sought vainly to get through from Hudson Bay to the Pacific Ocean. Brave attempts also were made to find the Northeast Passage—the route sailing eastward from the Pacific.

It was not until 1815, however, that the British Admiralty took up the search in earnest. In the following years, a number of expeditions were sent out under such intrepid men as Parry, Ross, Lyon and Franklin. By this time the existence of a passage had been proved, but it had little value because it was much too difficult for ordinary ships to navigate. The first successful crossing, by Amundsen in his ship Gjoa, took three years, from 1903 to 1906.

In any case, after 1860 the attention of explorers turned to the conquest of the North Pole. A number of American and British parties, under such leaders as Kane, Hayes, Hall, Nares and Greely, explored Ellesmere and other northern Arctic islands before Peary reached the North Pole in 1909 from a base on Ellesmere Island.

Exploration of the mainland of the territories was begun by Samuel Hearne, an employee of the Hudson's Bay Company stationed at Fort Churchill, on Hudson Bay. He crossed the Barrens—the far northern tundra—and reached the mouth of the Coppermine River in 1771–72, the first white man to reach the Canadian Arctic coast. Not long afterward North-West Company fur traders arrived at the Athabaska River (in present-day Alberta) and in 1789 Alexander Mackenzie descended the river named for him to the

PROSPECTOR NEAR WHITEHORSE
In a fast-running stream within a mile of the Alaska Highway, a prospector pans for gold.

Arctic Ocean, returning to his base at Fort Chipewyan after a journey of 102 days.

Trading posts were soon established along the newly discovered waterways, and for a time there was bitter rivalry between the North-West Company and the Hudson's Bay Company. The latter won out in 1821, and thereafter sent men to explore the Liard River and other tributaries of the Mackenzie system. They reached the Yukon River in the 1840's by way of both the Liard River (from the south) and the lower Mackenzie and Porcupine rivers (from the north). The coastline of the mainland was mapped between 1821 and 1853.

Whalers from Europe and the United States were among the first white men to penetrate the Arctic in search of a livelihood there. In fact, Herschel Island, near the mouth of the Mackenzie, was not only a fur-trading center but was also a favorite wintering place for whalers from 1890 until the decline of the whaling industry after 1906.

Originally the Northwest Territories included not only their present area but also the Yukon and northern parts of present-day provinces. The territories were under the British Crown; they did not become a part of Canada until 1870. It was even later, in 1912, that some of the area was surrendered to the provinces. Today the dividing line between the provinces and all the territories is the sixtieth parallel, stretching between Hudson Bay and the Alaska boundary. The Northwest Territories themselves are divided into three districts: Mackenzie (around the Mackenzie River system); Keewatin (facing Hudson Bay); and Franklin (the Arctic islands).

To go back a bit, in 1896 the discovery of gold in the gravel bars of the Klondike River (a branch of the Yukon) started a feverish rush to the Klondike. Some 100,000 persons set out for the Yukon and perhaps half that number actually reached the district. Many of them were ill equipped and perished from the hardships suffered on the way.

The next year the Yukon Territory was formed, with its eastern boundary marked off by the Mackenzie Mountains. On the map the territory has the shape of a triangle, wedged in between Alaska, British Columbia and the Northwest Territories. It has only a short strip of coast, on the Arctic Ocean. In 1899 Dawson City, then boasting 25,000 inhabitants, became

STANDARD OIL CO. (N. J.)

PADDLING ON THE GLASS-SMOOTH MACKENZIE RIVER
Four members of a scientific exploration party paddle light canoes on the Mackenzie. The great stream drains a massive area that extends south into Alberta and British Columbia.

THE ELDORADO MINES OF PORT RADIUM ON GREAT BEAR LAKE
Neat white buildings, tanks and stilted sheds vary the bleak rock shore of Great Bear Lake. Pitchblende from Eldorado supplies great quantities of uranium for atomic research.

the capital of the new territory. Agriculture and lumbering prospered, and a railway was built from Skagway, Alaska, to Whitehorse. More than $100,000,000 worth of gold was found in the first seven years after the Klondike rush. However, no new discoveries followed, most prospectors departed and the territory declined in importance. In 1921 barely four thousand people remained.

Before we consider what has been happening to the territories in recent years, let us see what the land itself is like. East of Great Slave and Great Bear lakes, the miles-wide Canadian Shield sweeps in an enormous semicircle, from the Arctic islands down to the Great Lakes and up around Hudson Bay. It is an exposed formation of Precambrian rock. In the hollows of the rolling surface lie thousands of lakes, many of them drained by the Kazan, Dubawnt, Thelon, Back, Coppermine and Anderson rivers. Steep cliffs, one thousand feet or more in height, occur where it edges water and at the western end. To the east, on Baffin and Ellesmere islands, it forms high mountain ranges. The thin soils of the Shield support only grasses, mosses, Arctic flowers, shrubs and dwarf trees, the trees becoming larger and more numerous as one nears the Mackenzie Lowland.

The Mackenzie Lowland is the northern portion of the great central plain of Canada and the United States. It is 600 miles wide at the sixtieth parallel but tapers to less than 100 miles at the Arctic Ocean. In the southern part are ranges of hills; and farther north, a mountain chain—the Franklin Mountains—lies between the Mackenzie River and Great Bear Lake. Most of the lowland is drained by the Mackenzie system which enters the territories at Fort Smith, as the Slave River, and after passing through Great Slave Lake emerges as the Mackenzie River. In places the Mackenzie is two to three miles wide, though before it reaches the Arctic it breaks up into many streams to form the Mackenzie delta. This mighty river of the north is fed by the Hay, Liard, Great Bear, Peel and Arctic Red rivers, and the system also drains two of the world's largest freshwater lakes, Great Bear and Great Slave. The most typical plants of the region (except in the delta) are poplar, spruce and birch trees.

West of the Mackenzie Lowland and extending to the Alaska border is the

NATIONAL FILM BOARD

PORTRAIT OF A THRIVING TOWN IN THE PIONEER NORTHLAND

Covering the promontory at the head of Yellowknife Bay and stretching into the lowlands beyond, Yellowknife has grown with the fast pace of industry in the Great Slave Lake region.

northward range of the Cordilleras, the backbone of North and South America. The Mackenzie, Selwyn, Ogilvie and Coast mountains continue the ranges of British Columbia, while the St. Elias Mountains extend into Canada southeast from Alaska. In these latter mountains is Mount Logan, 19,850 feet, the second highest peak in North America. The ranges stand on a high plateau which covers most of the Yukon Territory. Cut deep into the plateau are wide valleys in which flow the main tributaries of the Yukon River—the Lewes, Pelly, White and Stewart rivers—which drain most of the territory. The high altitude of the plateau makes it a grassland, but the valleys are well forested with spruce, poplar and birch.

In the Yukon Territory and the Mackenzie Valley, on the average, winters are long and cold but the short summers are surprisingly warm. However, the climate is erratic and may be mild or severe depending on whether gentle west winds blow from the Pacific or bitter winds blow from the Arctic or the interior of the continent. Snow covers the ground from October to March and frosts may occur in the summer months. Nevertheless the summers are long enough to allow some agriculture near Dawson City and along the Mackenzie, as the long hours of sunlight help to offset the briefness of the season.

The Arctic islands and the eastern mainland are always at the mercy of cold polar winds, which bring bitter winters and very short, cool summers. In fact, the Franklin District, or Eastern Arctic, lives up to the popular idea of the frozen north. In winter the temperature may remain below 20 or 30 degrees for weeks on end. Summer temperatures may go up as high as 70, and then hordes of flies and mosquitoes come out. The climate here is very dry. Rainfall averages only from nine to thirteen inches a year in the western part and from six to nine inches in the east and north.

White people form the majority of Yukon settlers; and Eskimos, 80 per cent of the population of Keewatin and Franklin districts. About half the inhabitants of the Mackenzie District (clustered chiefly along the Mackenzie waterway) are Indians, one-third are white and the rest are Eskimos.

The fur trade has continued to flourish in the territories, though on somewhat more prosaic lines. The more recent history of these lands centers about the growth of settlement, improvements in transportation and the gradual development of other natural resources besides furs. In the Yukon the growth of aviation and the building of the Alaska Highway during the war years revived settlement, though Dawson City has never recovered.

In the Northwest Territories the fur trade has expanded to include strings of posts along the west shore of Hudson Bay, along the Arctic coast of the mainland and among the northern islands. A number of mineral discoveries since 1920 have resulted in the development of an

FRONTIER BANK, YELLOWKNIFE
Bank of Toronto's log-walled branch serves the brisk financial needs of Yellowknife industry.

oil field at Norman Wells (the main source of petroleum for the Mackenzie District), one of the world's largest sources of uranium and radium at Port Radium on the eastern shore of Great Bear Lake, and a number of productive gold mines along the North Arm of Great Slave Lake. Here the town of Yellowknife has grown into the largest settlement in the Northwest Territories.

The war years increased the amount of shipping and improved transportation facilities along the Mackenzie waterway. Airports and weather stations were built. A six-hundred-mile pipeline—the Canol project—was constructed from Norman Wells to Whitehorse, though this was later abandoned. Since the war a highway has been laid from the Peace River area (of northern Alberta) to Great Slave Lake. There is now a commercial fishery on this lake as well.

Silky Pelts from the North

Many of the world's finest furs come from the territories, particularly Arctic fox, which provides almost the whole income of the Franklin District Eskimos, and muskrat, caught chiefly in the Mackenzie delta. Other fur-bearing animals abound—weasel, mink, red and cross fox, marten, lynx, wolf and beaver.

The trapping season for all except muskrat and beaver is the winter, when

PETROLEUM FROM THE SOUTH
Tanks lashed to a barge from Alberta contain Yellowknife's gas and oil for the coming winter.

AT POND INLET, BAFFIN ISLAND
An Eskimo and a hooded lad discuss the important business of hunting for seal and fishing.

the furs are at their prime. Snares are set along courses through the woods, and the trappers inspect them regularly. In the spring the trappers make their way to the trading posts to exchange the pelts for equipment, a few luxuries and some articles of food and clothing. In good years the income from furs has been large enough to enable some trappers to buy schooners and motorboats. Yet it is an uncertain livelihood. The trapper can never tell in advance how big his catch will be or what he will receive for it, as the price of furs fluctuates considerably. In bad years, traders, missions and the Government come to the rescue.

Fortunately, the trappers are able to obtain much of their food and clothing from hunting and fishing. Great quantities of fish are caught each year just to feed the sled dogs. Meat and clothing for the hunter and his family are provided by moose, bears, seals, white whales and, especially, the herds of caribou that migrate each year from the Arctic islands to the southern limits of the territories. Reindeer herds have been introduced and are established along the Arctic coast.

While most natives fish, hunt and trap, mining is the main employment of white settlers. In the Yukon Territory gold is still extracted from the gravel beds of streams. Copper, silver, lead and coal are mined, especially around Mayo and Keno Hill. The most exciting mining activity,

ON THE BARE NORTHERN TUNDRA—A FOREST OF ANTLERS
Reindeer, their fur molting but antlers in full glory, gather in a great herd at a government station, one of several in the territories reserved for native herdsmen and hunters.

NATIONAL FILM BOARD

SLOW GOING FOR A TRACTOR TRAIN ON ITS WAY TO YELLOWKNIFE
A plane drops weather reports and maps to a tractor train. The dangers of over-snow travel are many and great; sudden thaws and shifting drifts can maroon a train for days.

however, centers about the riches of the Northwest Territories, which we mentioned earlier. Hydroelectric plants have been built near Dawson City and Yellowknife to provide power for mining operations and for nearby settlements. Prospecting continues but is a far cry from the days of the Klondike rush. The modern prospector flies or drives and is likely to carry a Geiger counter.

Because of the climate and the distance from large markets, little agriculture is carried on in the territories, considering their vast size. Nevertheless, a number of farmers in the Yukon raise feed, livestock and vegetables and sell their produce and dairy products in the towns. Along the Slave, Liard and Mackenzie rivers large mission farms and many family truck gardens are cultivated. The farms in the southern sections produce much the same crops as the northern parts of the prairie provinces. North of Fort Simpson, however, agriculture is limited to hay, root crops and hardy garden vegetables. East of the Mackenzie and on the Arctic islands there are only experimental greenhouses.

In the Mackenzie Lowland and the valleys of the Yukon Territory, the forests more than fill the local needs for fuel and lumber. Elsewhere there is only driftwood and even that is very scarce.

The territories are, of course, directly under the Federal Government of Canada, which provides many services. Indians and Eskimos are helped by laws controlling trapping and the fur trade, and large areas are reserved for native trappers alone. Their standard of living is improved by Family Allowances as well as treaty money, and they receive free medical and dental care. Government doctors are stationed in the territories, and grants are made to the mission hospitals.

All settlers benefit from the work of many small Royal Canadian Mounted Police detachments, who patrol the most re-

THE ROOF OF NORTH AMERICA

NATIONAL MUSEUM OF CANADA

LOVELY NATIVES OF THE NORTH
The bell-like flowers and evergreen leaves of heather are a welcome sight on the tundra.

mote districts. Government radio stations supply a link with the outside world, and travel is helped by weather reports broadcast from observation posts.

Promising mineral discoveries are surveyed by government experts, who also watch over forests and farmlands.

Even the parts of the territories that have been fairly well known for more than a century are far from fully developed today. Undoubtedly more mineral wealth will be discovered. The growing importance of northern air routes is also helping to promote progress in the territories. Certainly the 25,000 people who live there today represent only a fraction of possible settlement. Yet until man has learned to adapt himself to a polar climate or there is more development, vast areas will remain empty, forbidding wastes.

By MORRIS ZASLOW

CANADIAN TERRITORIES: FACTS AND FIGURES

THE YUKON

Located north of British Columbia, this area, which was made a separate political unit in 1898, covers 205,346 square miles of land, but has a population of only about 9,000.

GOVERNMENT

The Yukon is administered by a commissioner (appointed by the governor general) and a Territorial Council of three members who are elected and serve for three years.

INDUSTRIES

Mining is the main occupation of the people; gold, silver and lead are the chief minerals. The region, however, is gradually increasing its production of timber, manufactured goods and fur pelts as it has an abundance of large and small fur-bearing animals.

COMMUNICATIONS

The territory has less than 60 miles of railway as most of its traffic is carried on the Yukon River and nearly 2,000 miles of road, including the Alaska Highway. Several commercial airlines provide passenger and freight services to Canadian and Alaskan cities.

EDUCATION

Besides the ten territorial schools conducted for white children in eight communities, there are educational facilities for Indians maintained by the Federal Government.

CHIEF TOWNS

Dawson, the former capital, population, 783; Whitehorse, the present capital, 2,548; and Mayo, 288.

THE NORTHWEST TERRITORIES

These territories include numerous islands and comprise 1,304,903 square miles. They have only about 16,000 inhabitants, most of whom are Indians and Eskimos.

GOVERNMENT

The region is divided into three districts: Mackenzie, Keewatin and Franklin; and their administration is vested in a commissioner, a deputy commissioner and an eight-member Council. Numerous Canadian–U. S. meteorological stations and air bases have been established on the most northern islands.

INDUSTRIES

Fishing and fur trading were the principal industries until recent years when gold, oil, pitchblende (ore of uranium and radium) and other minerals were discovered.

COMMUNICATIONS

Despite the all-weather Mackenzie Highway, boats and planes are the chief means of transportation, owing to the few vehicles and the long distances between the communities.

EDUCATION

Schools are operated in the main villages by the Federal Government and missionary organizations; and correspondence courses are prepared for children living too far away to attend the classes.

CHIEF TOWN

The area has only one town of any size, Yellowknife, in the Mackenzie District, with a population of 2,724.

The United States
How the Republic Spanned a Continent

As this chapter indicates, the physical geography of the land where forty-eight states are now welded together helped to shape its growth. In the early days, the long chain of the Appalachians, slanting southwest from Maine to Alabama, presented a formidable barrier. Once past that, the settlement of the Mississippi Valley was fairly easy. But beyond lay the towering Rockies, to which the Western pioneers paid untold sacrifice. Yet again the Americans were rewarded by finding lush green valleys on the Pacific coast. And so a great nation grew out of what had been a wilderness.

THOUGH the early settlers left their homelands in western Europe for a variety of reasons, once in the New World their first and all-important consideration was to establish homes. So, one after another, the little colonies were planted along the Atlantic seaboard—English in New England, Virginia and Maryland, Dutch in New York, many Germans in Pennsylvania, Scandinavians in Delaware and New Jersey, and French Huguenots in the Carolinas. In the first 150 years, only the Scotch-Irish newcomers were inclined to go any distance inland. They were the heralds of a long line of frontiersmen who were to push on ever farther toward the west.

Yet there was a certain amount of moving about within the eastern ribbon of settlement. Roger Williams, for instance, was banished from the Massachusetts Bay Colony because he disagreed with the all-powerful religious authorities. As a result, he and his followers established the Rhode Island colony.

Sometimes families were enticed to new territory because settlement gave a colony support for its claims on land. Thus Governor Van Twiller of New Netherlands lured a group from Massachusetts to settle in the Connecticut Valley, where there was much richer farm land than around Plymouth. Several other parties also journeyed from Massachusetts, notably one led by the Reverend Thomas Hooker, from Newtown, in 1636. This group drove 160 cattle through dense forests and swamps and across streams, carrying their household goods in packs on their backs.

Hooker's people made homes near the present cities of Hartford, Windsor and Wethersfield. Contrary to Dutch hopes, Connecticut was organized as an English colony.

The newcomers found no roads but only narrow trails worn by Indians, or by wild animals in search of salt licks. No journey was undertaken except for urgent reasons, and then it had to be on foot or on horseback or by canoe. For plodding through the thick woods, the pioneers quickly adopted the Indian moccasin. Snowshoes also came into use during the northern winters. Poles with wooden disks at the bottom—similar to ski poles—gave the traveler on snowshoes extra support. Even so, winter journeys were preferred. Then frost hardened the trails and ice covered the streams, and it was easier to slip through the woods. The forests were really dense in those days and were thickly tangled with vines.

Today, dog sleds call up a picture of the Arctic. Yet, for several years, early settlers in the northern colonies used them when the snow was soft or too deep for horses. Such a sled had a flat base of pine or spruce about two feet wide and curved up in front. It had room for only one person, and two to six dogs drew it.

Somewhat later a better kind of sled or sleigh was devised. If hauled by one horse, it was called a pod; if by two horses, a pung. The best highway for such a vehicle was the smooth surface of a frozen stream.

When a farm family journeyed forth to market in a sleigh, this was the picture:

In the bottom sat the farmer's wife and the children, all bundled up in coats, mufflers, hoods, mittens and blankets. Around them were heaped cheeses, dried herbs, hand-knit stockings and mittens, vegetables, and jugs of cider and vinegar. The man of the household trotted alongside the sleigh. To relieve hunger on the way, there was a large, round piece of frozen bean porridge. It might have been cooked days before, as the old nursery rhyme suggests:

> Bean porridge hot; bean porridge cold;
> Bean porridge in the pot, nine days old.

Early canoes were more like dugouts. A log was hollowed out by burning and chiseling with crude tools. The more typical canoe was made of birch, spruce or elm bark. Its frame, however, was of cedar or spruce; and the bark was attached to the frame by tough, slender larch or balsam roots. To prevent leaking, the seams were sealed with melted balsam or spruce pitch. The canoe was another idea borrowed from the Indians. Longfellow described it in *Hiawatha*:

> Thus the Birch Canoe was builded
> In the valley, by the river,
> In the bosom of the forest;
> And it floated on the river
> Like a yellow leaf in Autumn,
> Like a yellow water lily.

On smooth water, the canoe was an easy means of transportation. But falls, rapids or getting from one stream to another called for a portage. Then the canoe itself and any freight it held had to be carried around the obstacle or overland.

Travel by horseback was called "by post." A woman often rode on the same horse with her husband, sitting behind him on a pad, or pillion. When a woman rode alone, it was on a sidesaddle. A good horse was very expensive and might cost from $125 to $200. None could endure more hardship and strain than the Narraganset breed.

The early settlers made practically everything they needed at home—tools, soap, candles, homespun cloth dyed with home-made vegetable dyes. Money was extremely scarce, and the few things that

CHARLES PHELPS CUSHING

MASSASOIT—statue in Plymouth of the Indian chief who made a treaty with the Pilgrims.

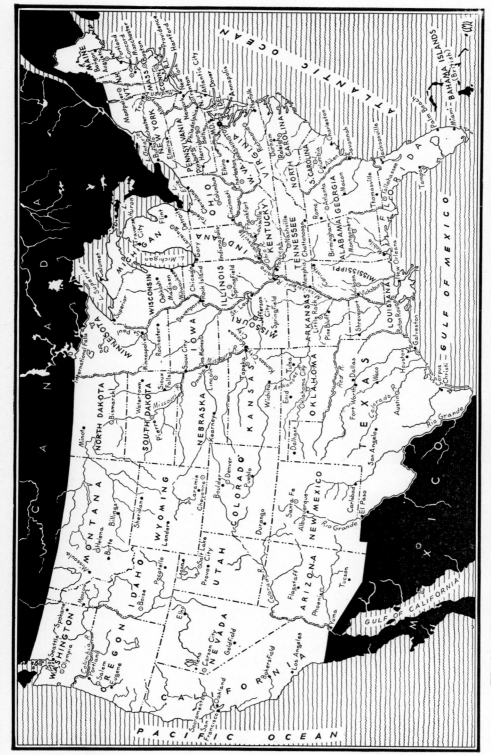

FROM ATLANTIC TO PACIFIC, FROM THE GREAT LAKES TO THE GULF OF MEXICO—ONE NATION

THE UNITED STATES

CHARLES PHELPS CUSHING

THE PURCHASE of Manhattan Island, carved on the base of a flag pole in New York.

dering over the country on foot and carrying his pack on his back. If he was fairly prosperous, however, he might have a pack horse. The animal was fitted out with a packsaddle and a pair of hobbles made of flexible twigs. Around its neck was a bell, silenced while traveling. At night the horse was hobbled and then the bell was unmuffled so that the horse could be located in the morning.

The First Highways

Road-making began when the Indian paths were broadened by hauling wider loads over them. Very early there were crude highways of this sort connecting neighboring villages. The first long stretch, however, was not opened until 1654. This was the Common Road between Boston and Providence, Rhode Island. Most famous of the colonial highways was the Boston Post Road, between Boston and New York.

The first road laws provided that they be ten feet wide with all trees cut close to the ground. As old accounts often speak of stumps in roads, the law seems to have been honored more in the breach than in the observance. In marshy or swampy places, corduroy roads were common— logs laid close together, crosswise over the highway. They made a solid but extremely bumpy surface.

Some dirt roads were of clay one to two feet deep. If it was wet, every time a struggling horse or ox pulled a foot out, there was a "pop" like the sound of a gun. Animals might become so deeply mired in the clay that they died there. The Bull-skin Road, for instance, which lay between Pickaway, Ohio, and Detroit, Michigan, was so named because of the many oxen that perished on it.

Vehicles did most of the actual work of leveling roads. First came clumsy two-wheeled carts. The wheels were cross sections of tree trunks, often six feet in diameter and six or more inches wide. Such wide wheels naturally helped to smooth a highway, and laws encouraged their use. If a man drove a cart with wheels six or more inches wide, he was exempt from road tax, and he did not

had to be bought—needles and pins, nails —were precious. To supply these articles, especially to isolated farms in the clearings, peddlers soon appeared. The peddler was a welcome visitor, not only for his stocks but also for the news he brought of the outside world. Though the Puritans of New England and the Quakers of Pennsylvania might frown on ribbons and trinkets, the peddler was more than willing to cater to the vanity of girls in other colonies. By the time of the Revolution, his stock was likely to consist of "Yankee notions"—tinware and brassware besides nails and pins—made largely in Connecticut.

The peddler was a solitary figure, wan-

CAPTAIN JOHN SMITH is remembered in a monument that looks out over the James River, Virginia. Near this spot he established the first permanent English colony in the New World.

FORT RALEIGH, on Roanoke Island, is today a national historic site. The blockhouses are reconstructions of those erected to defend the colony that disappeared so mysteriously.

BOTH PHOTOS, NORTH CAROLINA NEWS BUREAU

SCENE from *The Lost Colony,* an outdoor drama given every summer on Roanoke Island, off the North Carolina coast. It re-enacts the story of Sir Walter Raleigh's ill-fated settlement.

CHARLES PHELPS CUSHING

ELIHU COLEMAN HOUSE on Nantucket Island, south of Cape Cod, built in 1722. By that time prosperous New Englanders were erecting spacious though often austere-looking homes.

have to turn out when he met a cart with narrower wheels. Two-wheeled carts, however, were in general use but a few years.

The four-wheeled wagon is more typical of colonial days. Each of its two axles was a different length so that the front and back wheels would pass over a different part of the road. The tops of these wagons were covered with linsey-woolsey, a mixture of linen and wool cloth, or sometimes a mixture of nettle fiber and milkweed down. Under these tops women and children huddled in bad weather. The men usually walked.

Along frontier roads there were stations, or stores, only at long intervals, so food had to be carried for both human beings and animals. Bags of feed might be dropped along the way to provide food for the horses on the return trip. Human travelers carried large wallets filled with bread, jerked meat, boiled ham and cheese. The stations supplied mainly salt and sometimes a few nails, which were paid for in furs. Salt sold for $5.00 a bushel, and nails, from 15 to 25 cents a pound.

Until after 1776 practically no one traveled on Sunday. Indeed, in the Puritan settlements, no one might even walk in the streets on Sunday unless he was going to church. The law defined "Sunday" as from sunset on Saturday to sunset on Sunday. More than one traveler started on a journey Sunday evening, believing that the sun was down, only to have it appear from behind a cloud after he was on his way. For such a miscalculation, which seldom went unobserved, the traveler was fined.

The Puritans were no less strict in other matters. There was no place in their stern lives for play, even for the children. Football was to them "nothynge but beastlye furie and exstreme violence." Elsewhere in the colonies, however, social life was more relaxed. As anyone who has read *Rip Van Winkle* knows, the Dutch of New Netherlands loved bowling. The southern colonies, once they were well established, were gay. There was

PHILIP GENDREAU

A MINUTEMAN still stands in Lexington, Massachusetts, where the first shot of the Revolution rang out in 1775. Patriots were pledged to take arms "at a minute's notice."

FRAUNCES TAVERN, in New York City. It was a gathering place for prominent men during the Revolutionary period. Washington bade farewell to his officers here in 1783.

FANEUIL HALL, in Boston. Originally a public market, it was given to the city by a merchant, Peter Faneuil, in 1742. It served as headquarters of New England Revolutionaries.

SAMUEL CHAMBERLAIN

WITH A FIRE crackling on the hearth, the well-kept tavern was a comforting haven for travelers in wintry weather. The room is in the old Munroe Tavern, Lexington, Massachusetts.

much visiting among the plantations, and dancing was popular—minuets, quadrilles, reels. Christmas was celebrated joyously until Twelfth Night.

Children had a sweet tooth then as now, and it was amply satisfied in the seaports after trade began with the West Indies. Ships brought in not only sugar and molasses but also chocolate and ginger. One colonial shop sign read:

> I have Sucket, Surrip, Grene Ginger
> and Marmalade
> Bisket, Cumfet, and Carraways as
> fine as can be made.

Children living in the woods of New England feasted on maple sugar and syrup.

To the colonists the land seemed inexhaustible, and it was often used so carelessly that yields grew less. Even if they had known better how to conserve the soil, it seemed cheaper to move into new territory. They were not in the least daunted by the fact that backbreaking labor would be required to win another clearing in the forest. Thus settlement spread inland, and the frontier was pushed back to the foothills of the Appalachians.

For a long time rivers were the best highways, even though boats had to be propelled upstream against the current. A variety of craft were rigged with sails—pinks, pinnaces, ketches, schooners, lighters, shallops and pirogues. The pirogue, for instance, was really a very large canoe, often equipped with sails; it was sometimes forty to fifty feet long and six to eight feet deep. It could carry tons of household goods. If hostile Indians were encountered, however, a sailboat was practically defenseless.

As the colonies prospered and trade increased among them, a better means of hauling goods overland was needed than the clumsy wagons we have mentioned. The answer was the Conestoga, or Pennsylvania, wagon. It first appeared in 1755, although the name Conestoga was not bestowed on it until later. In that year, General Braddock used such a wagon on his expedition to Fort Duquesne

(Pittsburgh) in western Pennsylvania.

The Conestoga had a boat-shaped body that fitted it for mountain trails. No matter how the wagon was tilted, the cargo stayed in place. For feeding the horses there was a trough attached to the rear end. A Conestoga had six or seven bows —narrow, arch-shaped pieces—those in the center being a little lower than the end ones. These bows supported a covering of white canvas. The wagon was capable of carrying up to eight tons, though for each ton a horse had to be added to the team.

A wagoner was always proud of his fine-toned hame bells. In case he had trouble on rough or slippery roads and another driver helped him, it was the custom to give the hame bells to the rescuer. Sometimes a wagoner in trouble deliberately broke the tongue of his vehicle so that a passing driver could not help and, therefore, could not claim the bells. After the Revolution, paint made the Conestoga almost flauntingly patriotic. In contrast to the white of the canvas, wheels and sideboards were red, and the running gear and under part were blue.

The need for overnight stopping places brought about the establishment of inns or taverns. In fact, a Connecticut law of 1644 required each town in the colony to keep one. Americans used the word "tavern" more often than "inn," although the term "ordinary" was common in the South. Taverns were licensed, which gave a community some control in the selection of a keeper. Beer and wine were sold, and it would seem that the flavor of the wine left something to be desired, for it was the custom to spice it with nutmeg. Nutmegs were a luxury, and travelers carried their own in nutmeg holders made of wrought silver or Battersea enamel. A holder was just large enough to hold one nutmeg, and a pierced or corrugated surface on the inside of the cover served as a grater.

Taverns were often named for patriots, and above many a tavern door swung a crude portrait of Washington, Franklin, Pitt or Lafayette. Philadelphia had a

IN CRUDE HUTS such as these, the bedraggled Continental Army spent the bitter winter of 1777–78 at Valley Forge. Then the fate of the new nation was "suspended by a thread."

Four Alls Tavern before the Revolution. Its signboard read:

1. King—I govern all.
2. General—I fight for all.
3. Minister—I pray for all.
4. Laborer—I pay for all.

Even as restaurants do today, taverns frequently specialized in the food served. One tavern advertised fresh trout dinners; another served smoked ham that the owner himself had cured; still another served his own chickens. The waitresses were often the proprietor's daughters, and perhaps his wife—whose age had not "effaced the agreeableness of her features." Nor was the owner himself above giving service.

Early taverns took the place of newspapers or might have the only copy in a town. People went to the tavern to get information; and the proprietor himself was likely to be as inquisitive as any reporter. The story goes that when Ben-

PHILIP GENDREAU

BENJAMIN FRANKLIN arriving in Philadelphia—a famous statue by R. Tait McKenzie.

BLACK STAR

BETSY ROSS HOUSE, Philadelphia, where Mrs. Ross supposedly made the first flag.

THE LIBERTY BELL. From the steeple of Independence Hall, Philadelphia, July 1776, it rang out to proclaim the Declaration of Independence. The precious relic is inside the hall today.

THE HUDSON RIVER, one of the routes by which the early Americans first worked their way inland. A number of bridges span it now—here the Bear Mountain Bridge, near New York.

BOTH PHOTOS, NYXPIX-COMMERCE

VAN DEUSEN HOUSE, in Kingston, New York, is a charming example of the stone dwellings erected in the Catskill Mountains by refugee Huguenots from France, during the 1600's.

jamin Franklin entered a tavern, he would say:

> My name is Benjamin Franklin. I was born in Boston. I am a printer by profession, and am traveling to Philadelphia, shall have to return by ...... and have no news. Now what can you give me for dinner?

Any newspaper available was read until it was almost in shreds. A sign on a mantelpiece in one tavern read: "Gentlemen learning to spell are requested to use last week's newsletter." Before the Revolution, almost any kind of reading matter was at a premium. The few newspapers themselves were small and poorly printed, and books were scarce and dear.

The tavern was usually the social center of a town. On occasion, town meetings, religious services or theatrical performances might be held in it. If low-spreading trees grew nearby, the proprietor might build platforms connecting one tree with another. They made wonderful playhouses for children.

Outside of New England the theater began to flourish in the 1700's. The first colonial theater was built at Williamsburg, Virginia, in 1716; and by 1750 there were a number of Shakespearean companies. In 1766, the Southwark Theater in Philadelphia, said to be the first permanent theater erected in the colonies, opened with *Katharine and Petruchio*. (Evidently this title was preferred to *The Taming of the Shrew*.)

It was many years before Americans undertook lightly the hazards of an ocean voyage. Even in fair weather, it took a sailing vessel a month to cross the Atlantic by the northern route and two months by the southern. The experience of the famous Adams family gives us a glimpse of what a trip to Europe meant in the late 1700's. On one of his several diplomatic missions overseas, this time to Holland, John Adams took along his young sons, John Quincy and Charles. They sailed on November 13, 1779. The Atlantic was stormy, and the vessel leaked so badly that two pumps operated twenty-four hours a day to keep it from sinking. The ship was forced to land at Ferrol, Spain, and Adams and the children had to cross the Pyrenees Mountains in a carriage in wintertime. That part of the journey was perilous indeed, but they reached Holland without mishap.

A few years later, while Adams was still in Europe, his wife Abigail and daughter "Nabby" made the crossing. They sailed with two servants, John Brisler and Esther—who soon became Mrs. Brisler. The ship carried oil, which leaked, and potash, which smoked and fermented. Sleeping quarters were cramped, windowless and almost lacking in privacy. All four of the party were seasick. Moreover, the vessel was dirty, and this aroused the strong-minded Abigail's ire. She ordered brushes, mops and vinegar, and set Brisler to work. In a short time the ship was transformed, and the captain had the grace to thank Mrs. Adams.

Stagecoach Stations

As roads improved, stagecoaches became popular, especially after 1700. To accommodate them, there were stations every twelve or fifteen miles. At what was called a "swing" station, the stage stopped only long enough to change horses. At a "home" station, meals and lodgings might be obtained. A stage usually halted about ten o'clock in the evening and departed at three o'clock in the morning.

Stagecoaches were splashes of color—vivid red, blue, yellow or green. From four to six good, well-matched horses drew them. This was especially true of horses used on stages in the West. The stage driver was proud of his skill, independent and inclined to look down on other occupations. Stage driving, in fact, was such a good business that the same families furnished several generations of drivers.

Imagine what it must have been like to drive a team of spirited horses and a burdened coach over a deeply rutted road. If a stage approached a rut on the left, the driver would cry out, "To the right, gentlemen!" Then the passengers would lean to the right, sticking halfway out the stage windows. A moment later there might be a cry, "To the left, gentlemen!" So the stagecoach was right side up most of the time. On exceptionally bad roads, however, a stage might tip over or it might get stuck in the mud. Then the male passengers would generally get out and help the driver pull the vehicle out of the mudhole or set it upright.

One day, when a stage got stuck in the mud, the passengers were reluctant to help. Whereupon the driver sat down on a stone by the side of the road. After a few minutes a passenger asked the driver what he was waiting for. To which the driver retorted that he was waiting for the road to dry out!

Streams were always a problem to stagecoach drivers. Though some ferries could take a whole stage across, others could carry only the coach and the passengers. The horses had to swim.

To Charles Dickens, stagecoach travel in America was an ordeal, even though he experienced it at a later period. He wrote that the drivers chattered to each other at the stations like so many monkeys; and he detested their habit of chewing tobacco. Dickens was also displeased with the "chirping" of frogs and the grunting of pigs along the highway. As for the horses, he was quite sure that some of them were wild animals that had never been broken. Furthermore, he feared that riding over the corduroy roads would dislocate his bones. Of the few good features that Dickens admitted, one was that no driver could go to sleep because of the number of stumps in the roads. Nor could the horses run because the mud was so deep, and they did not have enough room to shy.

For all Dickens' strictures, stagecoaches were far superior to travel by post. The average person dressed well for a stagecoach trip, and he could usually count on arriving at his destination in a

VIRGINIA STATE CHAMBER OF COMMERCE

AN OLD-TIME CARRIAGE before the Governor's Palace in Williamsburg, Virginia. The palace was begun in 1705, burned in 1781 and restored in our century amid lovely gardens.

A CHARMING WAYSIDE INN—the Swan Tavern in historic Yorktown, Virginia. On either side of the door are hitching posts where guests once tied up their horses.

reasonably clean and tidy state.

During the colonial period, the "West" simply meant unsettled land, the habitat of Indians and wild beasts. By the time of the Revolution, however, the term meant the country between the Appalachians and the Mississippi River. Most famous, though not the first, of the frontiersmen who ventured across the mountains was Daniel Boone. Over the route that he blazed through the Cumberland Gap (later known as the Wilderness Road), settlers began to pass even before the roar of the Revolutionary battles had died away. So many new homes were established in the eastern half of the Mississippi Basin within the next few years that states were soon carved out of it: Kentucky, 1792; Tennessee, 1796; Ohio, 1803.

Life in the backwoods was rough, and out of it grew a special brand of American humor—the tall tale. Stories like the following were always told with an absolutely straight face. One day a man was riding by a swamp and noticed a large beaver hat lying on the surface with the crown upward. As he looked at it, the hat moved. So he touched it with his whip and underneath was a smiling head, which said, "Hello, stranger." The man asked if he could be of any help, but the head said, "No, thank you—I've a good horse under me."

No one could tell such tales better than Davy Crockett, who was elected to the Tennessee legislature largely on the strength of his storytelling skill. There is a host of hilarious legends associated with him, many of which he no doubt helped along. This man of racy wit was to die in the defense of the Alamo, in 1836.

When the United States purchased the Louisiana Territory in 1803, Americans knew little of the great tract beyond the Mississippi that they had bought. Nevertheless, there were some few white men already in the Far West. They were fur traders, trappers, hunters—a strange,

CUMBERLAND GAP. Through the break in the Appalachians, Daniel Boone blazed the Wilderness Road into Kentucky, in 1775. After the Revolution, it became a well-beaten highway.

wild breed called the "mountain men." Like Daniel Boone, they never had enough elbow room; and for the sweet sake of their solitary freedom they endured incredible hardships—sometimes almost starving or freezing to death. They abhorred and fled the civilizing influences brought by the advancing tide of settlement that began after the War of 1812. Most of them died unmourned and unsung. Yet it was they who made known the passes through the mountains and in many cases led the explorers and pioneers on the way west.

Lewis and Clark Expedition

When President Jefferson urged that the Louisiana Purchase be explored, Meriwether Lewis and William Clark were chosen to lead an expedition. Both of these young Virginians had been army officers and—a big asset—they were used to dealing with Indians. The expedition was ordered to go up the Missouri River as far as possible and to search out a route to the Pacific. It was also to report on the Indians of the region, animals, plants, minerals and trade possibilities. (Both Lewis and Clark wrote detailed journals, justly famous today, even with all their erratic spelling.) The permanent party consisted of the leaders, 23 soldiers (9 of these, skilled hunters), 2 interpreters and Clark's Negro servant.

On May 14, 1804, the group started up the Missouri River from St. Louis, in a keelboat and two pirogues. To promote friendly relations with the Indians, the explorers carried gifts—trinkets, medals, flags, red coats, paint and tobacco. The passage up the river was far from easy. A swift current had to be battled, there were snags and falling banks, and during the summer the men were plagued with "ticks, musquiters and knats" day and night. At the end of October, the party made camp for the winter in the Mandan Indian country (near present Bismarck, North Dakota).

Here the expedition secured the services of a Shoshone Indian girl as a guide for the coming trip through the Rockies. Sacajawea (the name means "bird woman") was the wife of a French Canadian trader, Toussaint Charbonneau, who also acted as a guide. (On February 11, 1805, Sacajawea gave birth to a son. The baby went along on the expedition all the way to the Pacific and back. Later Clark paid for the boy's education and he became one of the best interpreters in the West.)

In April 1805 the expedition pushed on. As it traveled west, it saw an abun-

THE HOMESTEAD near Reading, Pennsylvania, where Daniel Boone lived in his youth.

dance of wild game—buffalo, elk, deer, antelope, turkeys, magpies, prairie chickens, "barking Squerrels," beaver, trout, mountain rams, badgers and bears. Bears seemed to be especially troublesome. "These," Lewis wrote, "being so hard to die reather intimeadates us all; I must confess that I do not like the gentlemen and had rather fight two Indians than one bear."

The Great Falls in the Missouri River (in the present state of Montana) presented one of the worst obstacles. To make the portage around the falls and the rapids beyond, the men made rude carts, using the cross sections of tree trunks for wheels. On these carts the heavy canoes were piled, and horses hauled them. Though the portage was only eighteen miles long, it took a month.

THE UNITED STATES

ter was approaching again—a season of blinding blizzards and arctic cold in the mountains—so the party camped on the coast for several months.

Sea water was boiled to obtain salt; and a whale that the tide had washed ashore yielded meat and oil. One soldier wrote: "We mix it [whale] with our poor elk and find it eats well." During their stay the men killed 150 elk and 25 deer for the meat and skins. The elk skins were used to make 338 moccasins for the return trip.

On March 23, 1806, the group started on the long journey back home. It made good time and reached St. Louis on September 23. As many people had given the expedition up for lost, the men were welcomed with joy.

Venture into the Southwest

In the same year another army officer, Zebulon Montgomery Pike, led a party to locate the sources of the Arkansas and Red rivers. The explorers traveled as far west as Pueblo, Colorado, and gathered considerable information about the Southwest. On November 15, 1807, Pike sighted but did not succeed in climbing the mountain later named for him, Pikes Peak.

Although of less importance than the two expeditions we have just discussed, the one led by Major Stephen H. Long is in curious contrast. On June 6, 1820, the Long party started west from the Missouri River into Nebraska, on horseback. As it was summer, the explorers saw the thriving fields of corn, beans and pumpkins cultivated by Pawnee Indian squaws. Along the Platte River, large herds of buffalo and occasional bands of wild horses appeared. On June 13, the summit near Estes Park, Colorado, now named Longs Peak, loomed up. The odd part about this expedition is that Long, after exploring a considerable region, reported to the Government that the area between the Missouri River and the Rocky Mountains was unsuitable for farming. In his opinion its future value would be to keep the population of the United States from moving too far west!

PHILIP GENDREAU

PENNSYLVANIA "DUTCH" LANDSCAPE —barn built by descendants of German settlers.

All along the way the Indians were friendly, and this was probably due to the influence of Sacajawea. Through her efforts, the explorers were able to secure about thirty horses from the Shoshone Indians. Their chief was a brother of hers, whom she had not seen for five years. More horses were obtained from the Flathead Indians; and the Arikaras contributed corn, beans and dried squashes from their own precious stores.

In spite of this help, food became scarce. A long spell of rain hindered both exploring and hunting. For a time the men were forced to live on fish, dog meat and roots. Frequently they had to eat "portable soup" made from horse and dog meat.

At length they came to a river that proved to be the Columbia. They followed it to its mouth, and on November 7, 1805, Clark wrote: "Great joy in camp, we are in view of the Ocian." The expedition had achieved its main object. Win-

FRITZ HENLE, MONKMEYER

A "WIDOW'S WALK" surmounts an old house on Martha's Vineyard, an island off Cape Cod. In days of sail, a sea captain's wife kept watch for her husband's ship from the lookout.

THE PILLARED ENTRANCE to a lovely pre-Civil War home—Choctaw—in Natchez, Mississippi. In this city the chivalry and graciousness of the old South reached their fullest flower.

On his map, Long labeled the high-plains region the "Great American Desert." Thus it appeared in school geographies for half a century. Today the "Great American Desert" contains the largest irrigated area in the world and is one of the wealthiest agricultural and stock-raising regions in the United States.

Following in the wake of the government explorers, privately financed parties went west. Unfortunately, they were often less well equipped and suffered even greater hardship.

Between the years 1824 and 1830 Sylvester Pattie and his son, James Ohio, led a group of 116 men through the Southwest (most of which still belonged to Mexico). It was a record six years of endurance. Their horses' feet became so tender from being cut by sharp grass that they had to be shod with buffalo-skin moccasins. At one time the air was so hot that it seemed to burn the lungs, and the men could hardly talk because their tongues were swollen from thirst. To relieve it, they rolled bullets in their mouths, which would cause a little saliva to flow. When they finally reached water and began to drink, they became very ill. On another occasion, food was so short that they had to eat the horses. By 1827, only sixteen men remained with the company.

On their journey, the Pattie party stopped now and then to mine or to trap. A copper, gold, or silver mine could be rented from the Indians for five years for $1,000 a year. In copper mining, the Patties were fairly successful. As for trapping, once in a single morning thirty-seven beavers were caught. The company arrived in Santa Fe (New Mexico) with a valuable collection of furs, but the Mexican Governor seized them because

PHILIP GENDREAU

NEAR NEW ORLEANS, Louisiana, a typical Mississippi River ferryboat churns across the wide expanse. The great stream was, and still is, the chief gateway to the fertile Midwest.

THE PENNSYLVANIA MEMORIAL at Gettysburg. The terrible battle that occurred here on July 1–3, 1863, halted the Confederate forces and was the turning point of the Civil War.

GENERAL LEE'S HEADQUARTERS at Gettysburg. Though his army met defeat on the field, with dreadful loss of life, and continued to be harassed, it managed to make an orderly retreat.

the men had no trapping license from him.

From Santa Fe, the Patties continued westward, though it meant crossing the desert and Sylvester was ill. Eventually they reached San Diego, California, confident that they could secure food and water at one of the missions. Instead, the Mexican authorities arrested the Patties as spies and clapped them into prison. There Sylvester died. James gained his freedom some time later because he had been foresighted enough to carry smallpox vaccine with him. While he was in prison, a smallpox epidemic broke out in California; and he was released to give vaccinations. After all his years of effort, he got back to the United States a penniless man.

In 1834, Nathaniel J. Wyeth led twenty-one men, mostly farmers and artisans, to the Northwest. The journey began at Independence, Missouri, which was to become one of the assembly points for the emigrant trains taking the Oregon Trail. The Wyeth party followed the Platte River Valley, and the going proved so hard that about half of the men became discouraged and turned back. Only Wyeth and eleven others continued on to Fort Vancouver on the Columbia River. Among them was an ornithologist, John K. Townsend, who published his *Narrative of a Journey across the Rocky Mountains to the Columbia River* in 1839. The picture this book painted helped to fan the "Oregon fever" that now spread through the Midwest. Thousands started out from Independence, often to the tune of Stephen Foster's *O, Susanna*—"don't you cry for me"—played on a banjo.

On a later trip, Wyeth built Fort Hall in southeastern Idaho, which served as a gathering place and an outfitting point for emigrants for many years.

Ordeal of the Donner Party

The most tragic story of the western migration is that of the Donner party, which was among the first to include women and children. Two of the families were named Donner. In 1846 the group started for California, with high ideals of what it might accomplish. Mrs. George Donner carried books and paints along with which she hoped to start a girls' school. During the winter of 1846-47, the emigrants were trapped by snow in the Sierra Nevada, near Lake Truckee; and their sufferings were indescribably horrible. Valuable cattle were shot to end the beasts' misery and also to secure food—thirty-six of the animals had been left in the desert, some dead, some lost. When the last of their meat gave out, they ate the bark and twigs of trees, field mice and even their moccasins. When a stray deer was slain, they first drank the blood as it flowed from the wound and then ate the meat. Before rescue came, famine had reduced the company from eighty-one to forty-five; and the survivors were driven to cannibalism.

The Great Mormon Trek

Among the most successful treks westward was that of the Mormons. Their religious beliefs, especially the practice of polygamy, had aroused violent hostility in Illinois, and they hoped to find peace somewhere in the Far West. In the spring of 1847, under the leadership of Brigham Young, a mass migration began. It is estimated that altogether in that year 12,000 Mormons moved west, in companies of about 3,000 each. The march was brilliantly organized, with provision made for every possibility that could be imagined. A few leaders went ahead and established camps. They built log cabins, dug wells, plowed land and planted crops so that those who followed could depend on finding shelter and food. The last permanent camp was Winter Quarters, on the present site of Florence, Nebraska, just north of Omaha.

The route chosen lay along the North Platte River and through South Pass and Fort Bridger, both in Wyoming. In places, buffalo were so thick that the men had to drive them away to make room for the wagons to pass. With a device called an odometer, the leaders kept track of the number of miles traveled each day.

Eventually the advance party selected the Great Salt Lake area in Utah for a permanent home. When Brigham Young

A CONESTOGA WAGON, first designed for transportation over the Alleghenies. Its boat-shaped body was fitted for mountain roads. No matter how it tilted, cargo stayed in place.

saw the beautiful valley, he said, "This is the place."

Though the opening of the Far West was one of the most thrilling episodes in United States history, it must not be forgotten that during the same period the East was becoming ever more thickly populated. To provide easier transportation between markets on the Atlantic seaboard and the farms of the Great Lakes region, the Erie Canal was opened in 1825. It gave Midwestern agriculture a tremendous boost; and as other canals were built, traffic on all the waterways of the East increased enormously. Larger vessels were built, and travel by boat became popular. Then the Ohio and Mississippi rivers came into their own.

There was a great variety of river and canal craft—arks, barges, flatboats, keelboats and skiffs. Arks and flatboats were built of heavy timbers and often served as living quarters for the family of the boatman. Part of the upper deck was fenced off to provide play space for the children. Everything the family owned was carried along—furniture, farm implements, horses, pigs, cows, chickens, dogs, cats, kegs of powder, boxes filled with provisions. The boat was a floating log cabin, fort, barnyard and grocery store. The boat was laden with extra provisions to sell to settlements along the river.

The voyage downstream, in fair weather at least, was carefree. A boat anchored at whatever spot the family fancied. Toward evening someone usually played a fiddle so that the young people could dance. Who could resist a jig to *Turkey in the Straw?* Storytelling and singing also helped to pass the time on a trip that might last for months. To go from the mouth of the Ohio River to New Orleans took about six weeks; but a return journey, if it were made, required four and a half months.

Practically all the boats could be propelled upstream, but it took so much time and energy that most owners broke

NEBRASKA STATE HISTORICAL SOCIETY

PRAIRIE PIONEERS often lived in sod houses at first. There were no trees to provide logs, so turf was dug up and cut in the shape of bricks. Glass windows were a rare luxury.

PHILIP GENDREAU

THE PIONEER MOTHER, a statue in Kansas City, Missouri. The words around the base are from the Bible—Ruth's beautiful declaration to Naomi—"Whither thou goest, I will go . . ."

up their boats when they reached New Orleans or some other southern town. The timber might be sold or else the boatman would build himself a house with it. This practice was an important factor in the growth of settlement in the lower Mississippi Valley.

The bullboat was a western version of river craft. It was about 30 feet long, 20 feet wide and 12 feet deep, and had the shape of a brimless hat turned upside down. The frame was of willow, and this was covered with buffalo hides. A mixture of tallow and ashes calked the seams. Though a bullboat could carry six thousand pounds, it was so light that it could be taken to shore, turned upside down and used as a shelter. Lightness gave it another advantage because of the frequent portages that had to be made around the falls and rapids of western waterways.

Frontier boatmen had a stamp all their own. Usually tall, slender and sinewy, they wore a picturesque costume that set off their splendid physique—a bright red flannel shirt, a loose blue jerkin that extended to the hips, and coarse brown linsey-woolsey trousers. The boatman's hat was of untanned skins with the fur side out, and he was shod in moccasins. Attached to his belt was a hunting knife and a tobacco pouch.

A boatman's speech was no less colorful than his costume—an "iridescent vocabulary." Of a sudden occurrence, he said that it happened "quicker nor an alligator can chew a puppy." If he wanted a person to act quickly, he commanded, "Start yer trotters." Of a difficult task he proclaimed that it was "harder nor climbin' a saplin' heels uppard." To silence a long-winded individual, the phrase was "Shut off your chin music" or "Shut your mouth before you sunburn your teeth."

River navigation called for extreme alertness and skill. Often a rock large enough to wreck a boat lurked hidden beneath the water. A sawyer—a sunken tree bobbing up and down with the movement of the current—was another danger. Even worse was the sleeping (hidden) sawyer. Another hazard was the planter—a log firmly buried in a stream. As such obstructions were very difficult to remove, boatmen sometimes cut canals through islands.

As early as 1785, John Fitch believed that steam could be applied to navigation, and he invented a steamboat. However, the steamboat was a gradual development, and most authorities give Robert Fulton credit for the invention. At any rate, in 1807, his Clermont made the first successful trip in America, up the Hudson River.

Most early engines on steamboats

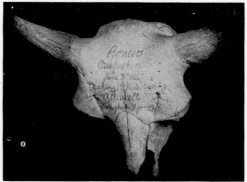

BUFFALO SKULL with an inscription by Brigham Young, leaving word of a camp.

A TRAIN OF MORMONS in the Utah territory. The picture is an early photograph, taken on July 10, 1878. Mormons continued to move west after the great migration of the 1840's.

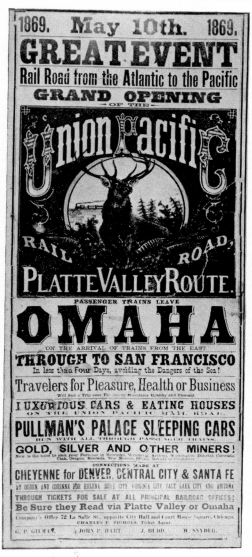

passion for racing each other. Then the steam pressure would be allowed to mount dangerously. Boiler metal was likely to be of poor quality, and it would burst under high pressure or from lack of water. In a race, a man might sit on the safety valve to keep up the pressure of the steam. So when the boiler exploded, he, and the passengers too, went up with it. Another perilous practice was to add pitch, oil, turpentine or lard to the fire to make it hotter and thus get more steam. Moreover, as the early steamboats were usually made of wood, they caught fire easily from sparks.

Life on the River Steamers

Accidents seem to have been an accepted hazard, for the river boats never lacked passengers; and they were often as colorful and oddly mixed a lot as might be found on an ocean liner—immigrants, migratory workers, patent-medicine vendors, theatrical troops, gamblers and preachers. On a boat, a preacher was sure of an audience, and he was equally certain that many of the passengers needed his services.

The Mississippi River pilots are immortalized in the works of Mark Twain, who was himself a licensed pilot. Such great skill was required of a pilot on this river that he might command a salary of $1,000 a month. He had complete charge of all the boat's movements except that he could not give orders on when to start or to stop. He had to know 1,500 miles of one of the most changeable rivers in the world. Obstructions were constantly shifting. High or low water, daylight or darkness—each altered the appearance of the stream. Besides, the very early pilot had no guide except eyesight and memory. Lighthouses and buoys were yet to be constructed, though navigation of the Mississippi would never be easy.

Pilot versus Captain

The boat captain took second place to the pilot, though the captain was often part owner of the vessel and was always its official master. It was a very foolish captain who dared to give orders to the

burned wood. Soon nearly every settler along the banks of the Mississippi River was chopping down trees and keeping a woodpile handy so that he might sell the fuel to the steamboat companies.

Practically all large steamboats had sleeping quarters, a dining room, a men's room and bar, and a ladies' parlor. The last was always at the stern of the vessel as that was the safest place when an explosion occurred. This happened all too often, because steamboat captains had a

GREAT DAY IN UTAH. On May 10, 1869, near Ogden, the last spike, a golden one, was driven in, completing the first transcontinental railroad. East and West were now linked.

pilot concerning navigation. The engineer, then as now, reigned supreme over the machinery.

Rough work on the boats was done by the roustabouts, most of whom were Negroes. They cleaned decks, stoked the furnace and loaded and unloaded the cargo. This group originated many folk songs—made up on the spur of the moment to lighten its labors. Any number of these songs are on record today, and those quoted below were published in the *Christian Science Monitor Magazine* from the collection made by Horace Reynolds.

As a Negro polished the brass in a pilot house, he sang:

> Wake up, Rose, put on your clothes;
> Day done broke, an' de sun done rose.

A roustabout who had hoped but failed to secure work sang this sad ditty:

> Boat all loaded
> And ready to back out,
> And the mate's going to leave
> This old roustebout.

One worker advertised his good nature and willingness to work with this:

> I loves everybody
> No matter how rough I seems,
> I wuk hard for muh livin'
> There's no use leavin' me.

Sometimes the verses consisted of only two lines:

> The prettiest girl I ever saw
> Lived on the banks of the Arkansas.

> Way down there in Arkansas
> The bullfrog kissed his brother-in-law.

Often a Negro would sing to show his pride in the boat on which he worked:

> De City of Cairo is a mighty fast one,
> But jus' lemme tell yuh what dat Monroe done,
> She left Natchez at half-past one,
> And landed at Vicksburg at the settin' of de sun.

Though the Negro contributed an important share to American humor, it was white men who brought it to public attention. Jim Crow Rice, the first of these men, roamed the rivers on showboats. He made up an act featuring the Negro stories and songs he had heard, painting his face black for the performance.

183

However, the first real minstrel shows were led by Dan (Daniel Decatur) Emmett. The appearance of his Virginia Minstrels in Boston in 1843 is regarded as the first genuine full-length minstrel show.

The Far West swung into the national limelight again with the discovery of gold at Sutter's Mill, near the present city of Sacramento, California. The lure of the yellow metal brought fortune hunters not only overland but also by way of the Isthmus of Panama and even all the way and horse broth. A number of horses had become exhausted and died along the banks. Those who passed through the Mormon community in Utah, however, were able to buy food from the successful farmers there.

The Panama route took very little longer than the overland trail, and the sea voyage to the isthmus was fairly comfortable. But many migrants came down with fever as they passed through the jungles of Panama. (It was yellow fever, and the cause was yet to be discovered.)

around South America. In the expectation of getting rich quick, they started out in high spirits. On the trail and on the ships, the strains of *Annie Laurie, Arkansas Traveler* and many another familiar song might be heard. Most of the singing of the overland group was done, however, before it reached Fort Hall, in Idaho. From there on, food and water ran short. Some of the weary travelers, wracked by hunger, ate soup made from their boots. Often they went for miles and miles without water. When they reached the Humbolt River (Nevada), its waters had the horrible taste of alkali

Once the gold seekers had reached the Pacific side of the isthmus, they often had to wait a long time for a ship. Vessels sailing north were dangerously overcrowded.

Long as the route around Cape Horn was, there were compensations on the voyage. If a vessel was at sea during the Thanksgiving season, there was a series of Thanksgiving dinners as each person on board insisted on keeping the Thanksgiving Day his state observed. It was not yet a national holiday.

On the heels of the gold rush, so many thousands of people settled in California

THE GODDESS OF LIBERTY lifts high her torch to light the way of those who seek freedom's shore. Bartholdi's famous statue in New York Harbor was a gift of the people of France.

INDEPENDENCE HALL, in Philadelphia, was built between 1732 and 1759 as the state house for the Province of Pennsylvania. The Continental Congress met here and voted the Declaration of Independence in 1776. Eleven years later, after independence had been won, delegates from twelve states framed the Constitution of the United States in the same room.

FOUR GREAT PRESIDENTS—Washington, Jefferson, Theodore Roosevelt and Lincoln—memorialized for the ages, carved in granite by Borglum on South Dakota's Mount Rushmore.

THE WASHINGTON MONUMENT rising high above any building dominates the city and has a thousand different aspects with all the varying conditions of time and weather. No view of it could be lovelier than this where we see it above the Japanese cherry trees which border the water of the Tidal Basin, and are reflected from it. The monument itself is over 555 feet in height, and is faced with white marble from Maryland. Though the cornerstone was laid in 1848 progress was slow, for one reason or another, and not until 1885 was the work completed

THE PAN-AMERICAN UNION, an association of the twenty-one republics of the Western Hemisphere, occupies what is undoubtedly one of the most beautiful public buildings in the world. It is built in Spanish, or Latin-American style, around a patio nearly sixty feet square, which is always filled with a luxuriant growth of semi-tropical plants. A glass roof covering the whole space is provided for the winter, but in the summer it is open to the sky. Below the cornice are the names of twelve men who are famous for their influence upon Pan-American history.

CHARLES PHELPS CUSHING

THE CAPITOL DOME in Washington gleams in the floodlights, as a shower of fireworks bursts nearby. The center of the building is of sandstone, the wings are of white marble.

that transportation became an urgent problem. The West now adopted forms that the East had used at an earlier period. Because distances were much greater, however, the West had more money invested in its transportation system than the East ever had during its stagecoach and freight-wagon days. One of the largest freight companies in the West was the firm of Russell, Major and Waddell. It transported military supplies along the Oregon Trail. Some idea of the scale of the company's operations is indicated by the fact that at the peak of its trail-freighting activity, it used 6,250 wagons, each with a capacity of 6,000 pounds, and 75,000 oxen. In 1859, Horace Greeley estimated that the property of this firm was worth $2,000,000.

Ben Holladay, however, was the greatest of the Far Western freighters and stagecoach operators. Between 1861 and 1866 he ran stagecoaches over a distance of 5,000 miles. He owned 500 coaches and express wagons, 500 freight wagons, 5,000 horses and mules and innumerable oxen. During his first year of business, the cost of equipment and of operating the lines amounted to $2,425,000. Holladay also ran steamship lines between Oregon, Panama, Japan and China.

Silver on Whip and Harness

Nearly every Western stagecoach driver worshiped his whip. He hated to lend it even to his best friend. Many of the whips had stalks ornamented with silver ferrules and silver caps. Harnesses also were adorned with silver, and a horse might have a silver bit worth $25 to $40. This lavish use of silver was considered a good investment because it advertised the mining industry of the West.

With the completion of the Union Pacific Railroad in 1869—joining East and West by track—stage lines began to go out of business.

Besides transportation, faster communication was needed between the West and East. The first mail received in California came by ocean steamer, and a letter from New York City to San Francisco was usually about four weeks in transit. Kit (Christopher) Carson, whose exploits as a trapper, guide and Indian fighter had made him a Western hero, carried the first mail overland from the Pacific coast to the East, in 1848. Ten years later the Butterfield Overland Mail started a more efficient service by stage, but it was still slow.

The upshot of all this was the Pony Express, a relay system that sped the mail through two thousand miles of an almost absolute wilderness, between St. Joseph, Missouri, and San Francisco. The first run started on April 8, 1860.

Pony Express Steeds

The "ponies" were really small, wiry horses. California mustangs were favorite steeds. Along the route, mounts were changed at about fifteen-mile intervals. Such a grueling pace was set that only young men, light in weight, physically strong and with iron nerves, were chosen as riders. Usually a rider was clad in a buckskin hunting shirt, cloth trousers tucked into high boots, and a jockey cap or a slouch hat, and a scarlet handkerchief was tied around his neck. A complete buckskin suit protected a rider in bad weather, for he was expected to keep going regardless of storms. Only two minutes were allowed in which to exchange horses and transfer the mail pouches at the relay stations. Buffalo Bill (William Cody, who organized his Wild West Show later) had the reputation of being one of the most able riders. It is claimed that on one occasion he rode 320 miles in 21 hours and 40 minutes.

One of the greatest enemies of the riders was the prairie dogs. A horse might step into one of their burrows and break a leg, throwing the rider and injuring him as well. Indians caused little trouble. During the nineteen months that the Pony Express was in operation, the Indians interfered seriously with the mail only once.

Each rider carried four small mail bags covered with oilskin silk to keep them dry. The total weight of any one mail could be no more than twenty pounds, and it usually weighed less. At first the

VIRGINIA CITY, Nevada, the most rip-roaring of all Western mining camps around 1880.

now moved westward to the rich timberlands of the Northwest—northern California, Washington and Oregon. Side by side with the lumberjacks paced their mythical hero, Paul Bunyan. If there was ever a real Paul Bunyan, who he was is lost in the mists of time. The giant lumberjack of legend, however, is one of the most amusing figures in American folklore. In the "dead-pan" style of the tall tale, he is credited with having dug about every deep crease in the surface of North America—the St. Lawrence River, the Grand Canyon, Puget Sound. For his Blue Ox, Babe, Paul gouged out the Great Lakes because Babe was always drinking the rivers dry. The Blue Ox was so huge that the width between its eyes measured twenty-four ax handles and a plug of tobacco. The mythical John Henry of Southern folklore was perhaps an adaptation of Paul Bunyan.

To return to the main thread of our story, while all the exciting activity we have discussed was going on in the Far West, back east the "iron horse" was coming to the fore. Though the Baltimore and Ohio, the first passenger railway in the United States, was begun as early as 1828, intense opposition to the railroads

Pony Express charged $5.00 for each ounce of mail or fraction thereof. A little later the rate was $1.00 a half ounce. In addition to the Pony Express charge for a letter, the regular United States Government postage of ten cents had to be paid.

On the average the Pony Express dashed from the Missouri River to San Francisco in 10½ days. The best time it ever made was when it carried the news of the election of Abraham Lincoln to the presidency—exactly 6 days, between the telegraph terminal at Fort Kearney, Nebraska, and Fort Churchill, in western Nevada. The end of the exciting Pony Express came abruptly, when the first telegraph line to California was completed on October 24, 1861.

In the preceding two decades, a "lumbering frontier" had opened in northern Wisconsin and Minnesota. This frontier

BOOT HILL GRAVEYARD, in Tombstone, Arizona. Desperadoes died with their boots on!

COURTESY LIFE MAGAZINE © TIME INC.

ON THE SALT FLATS in northwestern Utah, the wagon-wheel tracks of the Donner party are still visible. The little band passed this way before its grim ordeal in the Sierra Nevada.

WHERE SUTTER'S MILL stood on the American River, near Sacramento, California. Today the stream, once roiled by a host of prospectors in feverish haste, swirls by the site placidly.

BOTH PHOTOS, MONKMEYER

A TALE THAT IS TOLD. The discovery set off an extraordinary stampede. Lured by hopes of quick wealth, men rushed overland and by sea, even all the way around Cape Horn.

SAN JOSE MISSION in San Antonio, Texas, is called the Queen of the Missions for its architectural beauty. Founded 1720, it was restored in the 1930's as a national historic site.

CHARLES PHELPS CUSHING

BEACON ROCK rises 850 feet above the Columbia River, in Washington State. The tremendous monolith was so named by the men of the Lewis and Clark expedition, who used it as a landmark.

lingered for some time. In 1832, at Lancaster, Ohio, for example, an attempt was made to hold a meeting in the school for the promotion of a railroad. The school board replied:

"You are at liberty to use the schoolhouse to hold meetings for all proper purposes. But railroads and telegraphs are impossible and rank infidelity. If God had intended his intelligent creatures should travel at the frightful speed of 16 miles an hour by steam, he would clearly have foretold it in the holy prophets. It is a device of Satan to lead immortal souls down to Hell."

Steam Locomotives Appear

In the face of the railroads' obvious advantages, of course, opposition could not last. Even before locomotives appeared, there were crude trains drawn by horses on a track. Steam locomotives, however, took over in the 1830's. The first American locomotive designed for practical service was the Best Friend engine, built in New York City in 1830. But it was in use for only a short time when it exploded. In fact, train accidents were frequent for a number of years. Nevertheless, as engines were improved and safety measures came into effect, the general public was persuaded that travel by train—certainly much faster than by horse-drawn vehicles—was no longer fraught with peril.

At the same time, a ride on an early train was none too pleasant. The coaches were open, and passengers held spread umbrellas to protect them from the flying sparks from the engine. The cars looked as gay as a child's toy train. Locomotive and coaches were painted in vivid reds, blues and greens, and were lavishly ornamented with brass.

There were no timetables. A prospective passenger simply went to the depot and waited until a train came that was going in the direction he wanted. In places where one could see for a long distance, train officials sometimes erected a "lookout pole." A watcher climbed to the top of it and as soon as he saw any smoke, he dropped a note to the ground.

PHILIP GENDREAU

SACAJAWEA—a statue in Portland, Oregon, of the guide of the Lewis and Clark party.

Both railway and steamboat companies used a unique form of advertising. They pictured their trains and steamboats on Reward of Merit cards for children, evidently in the hope of catching the imagination of the coming generation of passengers.

At first each railroad builder used his own judgment as to the proper width of the track. The result was that there might be tracks of four or five different widths within one state. This meant that at the end of one line, passengers and freight had to be transferred to the next. Naturally the transfer consumed considerable time, and villages sprang up at the ends of lines to catch the trade the long waits brought. Passengers needed food and, often, lodgings. As a standard

THE UNITED STATES

gauge came into use, making the halts unnecessary, the folk who had come to rely on the trade sometimes tore new tracks up overnight.

The Indians also tore up many railroad tracks because they knew that the railroad would bring white settlement. They had not forgotten that the white man had almost killed off the buffalo, depriving the Indians of clothing, food and shelter. While the Union Pacific Railroad was being constructed, through wild country, the Government had to post soldiers along the roadbed to protect the track.

Today Americans are likely to take railroads and all the other swift means of transportation and communication for granted. They can fly overnight to Europe or South America seated in a warm cabin, or talk with casual ease to a friend halfway around the globe. Yet all this is but the climax to an extraordinary story of development, which paced the whole history of the United States. Tracing the beginnings of the story and what it meant in human effort yields greater understanding of the modern nation.

By Ora Brooks Peake

THE UNITED STATES: FACTS AND FIGURES

THE COUNTRY

The total area of continental United States is 3,022,387 square miles (land area 2,974,726 square miles); population as shown by 1950 census, 150,697,361.

GOVERNMENT

Federal republic of 48 states; executive power vested in a president and a cabinet of 10 members; legislative power in a Congress consisting of 2 houses, a Senate composed of 2 members from each state and a House of Representatives of 435 members (each state is given representation according to its population). Each state, except Nebraska, has a legislature of 2 houses, a governor and other executive officials. Both houses of the legislature are elective, but the senators have larger electoral districts.

COMMERCE AND INDUSTRY

The United States ranks first in world production of corn and cotton; other important crops are wheat, hay and forage, vegetables, fruits, oats, rye, barley, buckwheat, rice, tobacco and sugar crops. Swine, sheep and cattle are raised in great numbers; 209,671,000 head of livestock of all kinds (1949 est.). Mineral resources are rich and varied; of the metals, iron ore is first in value, followed by copper, zinc, lead, gold, ferroalloys, silver, tungsten, bauxite, mercury, platinum. The non-metals include coal, petroleum, natural gas and a variety of building materials. Petroleum is the most valuable mineral product. The chief timbers are fir, pine, hemlock, spruce and redwood. Extensive salt- and fresh-water fisheries. Manufacturing has become highly developed; some of the important industries are motor vehicle manufacture, slaughtering and meat-packing, iron and steel, steel works and rolling mills, petroleum refining, printing and publishing, flour mills, textile mills and electric light and power generation. Chief exports are raw cotton, petroleum and its products, industrial and agricultural machinery, automobiles and accessories, grains and grain products, iron and steel products, tobacco, cotton manufactures, lumber products; leading imports are wool, rubber, coffee, sugar-cane, paper and paper manufactures, cocoa, petroleum, hides, skins, furs and copper.

COMMUNICATIONS

Railway mileage, about 225,000; total length of telegraph wire, 1,748,000; ocean cable, 103,671 nautical miles; length of telephone wire, 167,700,000 (1952). There were 2,600 commercial radio broadcasting stations in 1953. Commercial air transportation has had rapid development since the passage of the air commerce act in 1926. 6,042 municipal, commercial, government and private airports in operation; 77,349 miles of domestic airways and 110,213 miles of international airways in operation.

RELIGION AND EDUCATION

No established church; all denominations represented. Protestants about 59 per cent of church membership. Each state has a system of free public primary and secondary schools supported by state and local taxation. There were in 1952, 31,374,000 pupils enrolled in elementary and secondary schools (public and private). There are 1,851 colleges and universities including junior colleges and professional schools.

CITIES

Washington, the capital, in District of Columbia. Population (1950 census), 802,178. For other cities, see Sectional Summaries.

OUTLYING TERRITORIES AND POSSESSIONS

Alaska Territory, Hawaii Territory, Puerto Rico, Guam, American Samoa, Panama Canal Zone and the Virgin Islands. The total area of these possessions, 581,702 sq. mi. Population (1950), 3,535,873.

FROM MAINE TO MARYLAND

Town and Countryside in the Historic Northeast

The Northeast is not a geographical unit, and the territory from the Potomac to Canada possesses fascinating variety. New Hampshire is different from Delaware, Boston is not like New York City, yet every one of the eleven states has some interests in common with the others, and the charm of Colonial times lingers here and there in them all. This is a region rich in the beauty of shore and mountain and rolling countryside, rich in the resources of mines and water power, with great industrial and commercial cities and many centres of education. Altogether, the long-established states at the eastern gateway of the continent form one of the most important, populous and beautiful sections of the whole country.

NO part of the United States is more richly interesting than the Northeast. Historically, scenically, industrially—whatever the viewpoint—the states of the Atlantic seaboard from Maryland and Delaware north to the Canadian border are a fascinating group. Here history was made in Colonial and Revolutionary days, and every changing development of the nation has been reflected in this region. New England was the first industrial section in America and is still one of the great shoe and textile centres. Important manufacturing cities dot the map from Maine to Maryland, and farms of many kinds help feed the millions engaged in commerce and industry. The countryside is extremely varied and each section beautiful in its own way, whether we like best the flat salt marshes of the Jersey shore, the Finger Lakes of western New York or the stony elm-shaded pastures and the rocky coast of New England.

These contrasts in the landscape show us vividly that the different states are not the same geographically, and that the state boundaries do not follow natural physical lines. Of all the eleven northeastern states, Delaware is most nearly a unit; except for a small section near its northern boundary, it belongs entirely to the Coastal Plain which, as we remember, stretches from Texas and Florida to Long Island, and includes the low sandy point of Cape Cod. Half of Maryland and much of New Jersey are thus part of the Plain. Back of it in Maryland, Pennsylvania and northern Jersey are the rolling hills of the Piedmont Belt, and then come the Appalachian Ranges—the mountain backbone of eastern North America. The rough plateau country west of the Appalachians covers all western Pennsylvania and most of southern New York. The Adirondacks are separated from the mountains of northern New England by the valleys of the Hudson and Lake Champlain. Vermont is almost entirely mountainous and so is half of New Hampshire, but the coastal region is lower.

Of all the Northeast, New England was the least inviting when the first colonists from Europe came looking for new homes, but it was settled nevertheless, and its people early took a leading part in American development. The various colonies at Plymouth, Massachusetts Bay, Providence, New Haven and elsewhere gradually came to be connected by more settlements, and a certain sort of unity was forced upon them by the necessity of fighting off French and Indian attacks. It was a hard life they led, and only settlers strong in body and spirit could survive; but they prospered in spite of Indian wars, severe winters, and in places a thin or patchy soil. They made their own clothing, shoes and furniture, because they had to, and the many hand industries which developed at this time were one basis of later manufacturing supremacy. The colonists also had an interest in education, largely due to their deep respect for Biblical and theological learning. Thus it happens that New England can boast the first factories, the first printing-press and the first college in the United States.

WARREN, MARYLAND DEPARTMENT OF INFORMATION

THE GREAT CURVE of the Chesapeake Bay Bridge. It is seven and one half miles long and crosses the bay south of Baltimore, linking the western and eastern counties of Maryland.

U. S. NAVY

THEIR WHITE UNIFORMS blazing in the sun, midshipmen march in perfect formation across Tecumseh Court, in front of Bancroft Hall, at the Naval Academy in Annapolis, Maryland.

BAUER, FROM CUSHING

ROUND TOWER or Old Stone Mill—Newport, Rhode Island, landmark. Whether it is an eleventh-century Norse relic or a seventeenth-century mill remains a mystery to this day.

New York, of course, was settled by the Dutch, and did not come into English hands until 1664, when upper New Jersey also became English. In Delaware the Swedes had been first, but the Dutch governor of New Netherlands, sturdy old Peter Stuyvesant, conquered New Sweden and held it until he himself was expelled by the English. Maryland was the first to insist upon religious freedom. Pennsylvania, also famous for its toleration, was not founded until 1682, but it grew rapidly and attracted colonists from the British Isles and Germany.

Just as New Englanders were forced at an early date to join together for mutual protection, so the whole group of English colonies learned a certain sort of co-operation in the almost continuous fighting against the French and the Indians. Yet such was the self-reliance and independence necessarily developed by each colony that it is a wonder they ever stuck together firmly enough to carry through the Revolutionary War. Boston was from the first a center of resistance, and Philadelphia was the meeting-place of the Continental Congress and the Constitutional Convention. Everywhere throughout the Northeast are reminders of Colonial and Revolutionary days. The number of houses where Washington slept or made his headquarters has become proverbial.

We are less likely to find monuments commemorating important events of the industrial revolution which followed upon the political one. Mechanical and economic changes of far-reaching importance coincided roughly with westward expansion. At the beginning of the nineteenth century the United States was a nation of small farms and hand industries; small cities and many farms were scattered along the seaboard north of Florida, and a few outposts stood in the wilderness between the Appalachians and the Mississippi. By the early twentieth century, the territorial boundaries had reached the Pacific and extended overseas, and the nation was largely urban and in-

STANDARD OIL CO. (N. J.)

THE STATE OF POTATOES

Many hours of backbreaking labor are eliminated by this tractor-drawn machine that is shown digging up potatoes on a farm in Caribou, Maine. In spite of the fact that only 33 per cent of the total acreage of Maine is suitable for agricultural purposes, it exceeds by many thousands of bushels the potato crop of any other state in the Union.

FRESH OYSTERS from the beds planted in Long Island Sound are prepared for market.

dustrial, with manufacturing done in factories on a gigantic scale. Water power, steam power and finally electric power—the development of such resources accompanied and made possible the invention of complicated and delicate machinery which took industries out of the home and required large capital resources.

The Northeast experienced these industrial changes first. In New England, especially, thousands went into the factories as farming grew less and less profitable, while other thousands went west to settle new states. Hardy spirits took to the sea in greater numbers than before, as fishermen, whalers and sealers, and fast clipper ships from Boston or Baltimore did much of the carrying trade of the world until steam-driven ships of iron destroyed their supremacy. Fresh immigration kept adding to the cities, so that the racial make-up of the population had entirely changed, and newcomers from Europe and their descendants outnumbered the descendants of the original colonists. In New York City the results of immigration and industrialization were

most pronounced of all, on account of its commanding location and its fine harbor, but Philadelphia and Baltimore also grew to be great seaports and railway terminals, and Pennsylvania coal and coke made Pittsburgh into a center of the iron and steel industry.

Thus the wilderness of 1620 was transformed in three centuries. Could the Puritan leaders see Boston today—could Peter Stuyvesant see Manhattan Island, or Penn his Philadelphia—they would think themselves in another New World. Yet if we travel from end to end of the Northeast we shall find the transformation not always complete. There are stretches of countryside in New England which still look much as they did in Colonial days; there are old farmhouses and meeting-houses in Pennsylvania and New Jersey which date almost from Penn's time, and boxwood grown from cuttings of English shrubs still grows on old estates of the Maryland Eastern Shore. Such survivals of an earlier day give the East much of its beauty.

Those who have not seen New England

PHOTOS, CONNECTICUT DEVELOPMENT COMMISSION

THE QUIET LIFE of a small New England town is pictured in Farmington, Connecticut.

Courtesy Boston and Maine Railroad

THE SPIRIT OF NEW ENGLAND PERVADES THE OLD MANSE AT CONCORD

No place in New England is more beautiful and more memorable than Concord, Massachusetts, and the Old Manse with its gambrel roof shaded by a gracious elm tree is one of Concord's most famous houses. Here Emerson once lived and here Hawthorne wrote Mosses from an Old Manse. The bridge defended by the minute men is not far from the house.

Courtesy Boston and Maine Railroad

CRAIGIE HOUSE BELONGS WITH THE TRADITIONS OF CAMBRIDGE

The stately grace of Colonial mansions is in every line of beautiful old Craigie House. Washington made it his headquarters in the early part of the Revolution, and later it was the home of Henry Wadsworth Longfellow. It stands on Brattle Street and faces Longfellow Court, a lovely little park named in honor of the poet.

Photograph by E. G. Wooster
ONE OF NEW ENGLAND'S MAGNIFICENT ELM TREES
Without the gracefully arching elms which shade every village green, New England would be infinitely poorer. A fine tree can lend beauty to the commonest country road, and give character to any house. This white farmhouse at Ridgefield, Connecticut, set in a bend of the road under its great tree and surrounded by its picket fence, is typical of the countryside.

Photograph by Leonard Schwartz
THE CHARM OF THE COLONIAL LINGERS ON IN DOVER
There is no place in the United States where the atmosphere of former days persists more delightfully than in the capital of Delaware. Dover was founded in 1717, and the Ridgely house, built of quaint red and black-faced bricks brought from England, is one of the beautiful old houses which face the quiet Green in front of the State House.

cannot realize its charm. Compared with western North America or South America its scenery is not at all spectacular; here are no stupendous peaks fifteen or twenty thousand feet high, no snow-covered volcanic cones, no enormous glaciers. The low wooded mountains, the rocky fields, the pockets of good soil, the many lakes large and small tell the geological story of an old land, with hills eroded and scarred by glacial action. Valuable granite and marble quarries are concealed in those hills, and the beautiful lakes to-day provide ideal locations for many a summer camp, in Maine, New Hampshire and Vermont. Spruce and pine, hickory and hemlock still cover large areas, fringe the tortuous coast of Maine, and seem to summer visitors like the forest primeval, but most of it is second and third if not fourth growth. Lumbering is less and less profitable in the Northeast and much reforestation is necessary. The tourist business has become, in some parts of New England, even more important than lumbering or farming. Farmers take summer boarders or rent summer camps; old inns become famous, new hotels are built and the resort trade of three months supports many a community for the balance of the year. Along the cool Maine coast, Mt. Desert, Bar Harbor, Penobscot Bay and Boothbay Harbor are names known alike to tourists, week-enders, tired vacationists and fashionable summer colonists. The combination of surf and rock and forest has a strong appeal for many people while others prefer the deeper woods of the interior, where long and exciting canoe trips lead from stream to stream and lake to lake.

Hills and Mountains

For those who like the hills better than the sea, Vermont has its Green Mountains and New Hampshire its White Mountains. Both ranges are green in summer and white in winter, and the call of winter sports in this section is now almost as insistent as the summer lure of mountain villages, forest trails and blue lakes hidden away in the green hills. Massachusetts in turn has the lovely Berkshire Hills, and in addition a coastline of superb variety. The North Shore—that is, the shore above Boston—boasts a succession of beautiful headlands and wholly delightful old towns: Marblehead, fascinating Salem, Beverly, and Gloucester in the lee of Cape Ann. This is the region of fishing-schooner and clipper-ship fame; sails are still mended on Gloucester docks and codfish spread out in the drying yards. Gloucester and Boston ship cod, herring and mackerel even to distant points, but those who have never eaten fresh Gloucester mackerel or swordfish new-caught off Martha's Vineyard do not realize how delicious fish can be.

Plymouth and Cape Cod

Then there is the South Shore, past Quincy and Hingham to Plymouth and Cape Cod. "The Cape" is known to thousands who love its sandy roads and high dunes, its beaches and its tough, wind-bent pine and juniper trees, its weathered old frame houses with dooryards of asters, and its white picket fences lined with tall dahlias. Nantucket and Martha's Vineyard, those two low islands across Nantucket Sound from Chatham and Falmouth, are akin to the Cape geologically; they were settled by the same kind of sea-faring folk, who went on fishing and whaling voyages and built the same kind of low charming houses.

Little Rhode Island

The greatest whaling port of all was New Bedford, as readers of Moby Dick well remember. Then it became one of the great cotton textile towns, like Fall River just to the west and Pawtucket and Providence across the line in Rhode Island. The smallest state in the Union has almost more water than land, for Narrangansett Bay divides it in two, cutting deeply inland. The low hills of Rhode Island and Connecticut are much alike in character, and are beautiful with many swift-moving streams whose waterfalls—as in the rest of New England—have determined the locations of factories. A full description of New England's manufacturing towns is here impossible.

ATLANTIC CITY, POPULAR PLAYGROUND OF THE NEW JERSEY COAST
The boardwalk and the wide, sandy beach make Atlantic City a Mecca for vacationers. The boardwalk is lined with luxury hotels, shops, restaurants, a convention hall and movie theaters.

COURTESY GREEN MOUNTAIN CARD COMPANY

ON COLD DAYS in early spring when the sap begins to rise it is time for the first step in the making of New England's sweetest product, and the Vermont farmer sets out to tap his sugar maples. He bores a hole in each tree and places a small bucket under each tap hole. Later, as we see here, he makes the rounds again to get the sap which has collected. The barrelful on this sled will be added to others at the sugar house and then will be boiled down to make the delicious clear maple syrup and cakes of brown toothsome sugar for which Vermont is famous.

MULBERRY STREET on New York's lower East Side has undergone many changes since this colorful photograph was taken. But modern housing projects have not completely replaced the slums, and tenement buildings like these still stand throughout the area. The children living in today's cramped tenements need not play in the busy streets. They can usually find city playgrounds near their homes or good recreational facilities in community houses. Many of the children spend their summers in country camps, sponsored by various welfare organizations.

There are the huge paper mills of Millinocket in the heart of the Maine woods, the woolen and cotton mills in the cities of Lowell and Lawrence, the shoe factories of Haverhill, Lynn and Brockton; watches come from Waltham and clocks from Waterbury, jewelry from Attleboro and typewriters from Hartford. It is surprising how many articles of every-day use are made in this small area.

The Effect of the Lakes

New York in turn presents a fascinating combination of scenery and industry. The Adirondacks are more varied than the Green Mountains, from a geographical point of view. Lake Ontario and Lake Erie influence the climate and the products of western New York favorably, so that it is a country of orchards and vineyards, and in good years fruit has been so plentiful as to go unpicked. Throughout the Northeast, farming must be intensive and specialized or it is profitless, and abandoned farms, whether in Vermont, New York or Pennsylvania, tell the same story of non-adaptation to economic and local conditions.

The Route to the West

It is the Barge Canal and the railroads following its route which link east to west in New York. Add cheap transportation to hydro-electric power, and the result is a string of manufacturing cities from Buffalo to Albany, each well known in its line: optical and photographic goods from Rochester, collars and cuffs from Troy, electrical machinery from Schenectady. Down the Hudson Valley from Albany pours the volume of commerce which determines New York City's leadership in trade. On each bank is a railroad, and the river itself is a highway. It is one of the country's most beautiful streams, cutting its way down from the heart of the Adirondacks, joined by the Mohawk and flowing on past the Catskills through the highlands until it spreads out at Tappan Zee into a bay four miles across; then suddenly just below Nyack the Palisades begin, and those magnificent cliffs line the west side of the river until we are opposite the Manhattan skyscrapers. The counties along the Hudson were settled by the Dutch, and place-names from Staten Island to Rensselaer are echoes of the days when New York was New Netherland, and stocky, vigorous settlers built white frame houses with gambrel roofs in the style which we call Dutch Colonial.

No description of the state, however brief, could leave out Long Island, with its beautiful estates and towns, and fertile truck gardens growing produce for city markets. There is a pleasant and comfortable atmosphere which often gathers about fields well cultivated and clusters of communities, and these Long Island has in good measure. The long sand beaches and dunes proclaim the island's relationship with the Jersey coast to the south.

The Surface of New Jersey

One-third of New Jersey is hilly country with parallel mountain ridges walling pleasant valleys and well-kept dairy farms. Factory towns like Paterson and Passaic, with their silk mills, half encircle New York City. But from the Raritan south the land is relatively flat, and literally so on the beaches which barricade the coast. Southern Jersey is all sand, as though it had been raised from under the sea only yesterday, and such, geologically speaking, is the case. How can anything grow in that sandy soil? Many things, such as grass and grain crops, will not, but melons, potatoes, fruits and truck crops will. So will stubborn pines and scrub oaks. Thus the central Jersey landscape is one of flat fields carefully fertilized, cultivated and watered. A few country towns, farms and fine old homes still stand as reminders of the colonial past. But the countryside is steadily giving way to industrial communities and hundreds of acres of housing development. Farther south are wide stretches of pine woodland where white sandy trails show up distinctly against the dark green foliage. The Jersey which most people know is the shoreline itself, that series of beaches each isolated by its bay or inlet and by wide, desolate salt marshes. On the beaches the Atlantic breakers pound steadily, and there is good surf bathing by the mile. Consequently this coast is practically one long summer resort. Asbury Park, Seagirt, Barnegat, Beach Haven, Atlantic City, Ocean City, Wildwood, Cape May—every city dweller in the near-by metropolitan districts knows one or all of them. In fact so many people run away from the suffocating heat of Philadelphia and New York that the barrier islands are almost entirely built up with cottages, hotels and board-walks and there are few of the big sand-dunes left.

Away from the sea breezes and the fishy salt smell of inlets, once across the flat truck lands and, finally, the Delaware River, the different character of Pennsylvania landscape is immediately evident. Southeastern Pennsylvania is rolling, hilly, Piedmont country of good soil and many streams. It has been well cultivated for two hundred years, and more than one family still holds land by deeds from William Penn. The Quaker settlers built houses and barnes of local sandstone, and developed an architecture known as the Pennsylvania farmhouse type, different from the New England and Dutch Colonial homes, but quite as satisfying to look at. North and west is the belt of country settled by Germans who are often called "Pennsylvania Dutch." These people long kept their racial identity, their customs and their German dialect almost intact. Big stone houses and great red barns painted with white circles and stars are characteristic of the region, which stretches around in a wide quarter-circle from the Maryland line through the tobacco fields of Lancaster County to Reading, Bethlehem and the Delaware.

BLACK STAR

WEST POINT cadets march past the ivy-covered gray walls of the Military Academy on their way to the parade ground.

River transportation has helped to concentrate a ring of industrial cities around Philadelphia, and is especially responsible for the shipyards of Camden and Chester, while blast furnaces smoke up the valley of the Schuylkill which leads down from the hard coal fields. One of the country's greatest steel plants is located at South Bethlehem, on the edge of the Appalachian ridge country, where Lehigh Valley coal is easily obtained. Much cement is manufactured in this limestone country, while the biggest slate quarry in the world is near Bangor, above Easton.

Almost all the anthracite coal in the United States is in one relatively small

THE DEVIL'S PULPIT at the jagged tip of Bald Head Cliff is near York, on the coast of southern Maine. The half-severed cliff looks as though someone had tried to chop it in two with a giant butcher's cleaver and resembles the steep rock wall of the side of a quarry from which great slabs have been cut. There is a strange fascination in the ebb and flow of the water, as hour after hour the blue-green waves roll up ceaselessly, curl, and break foaming on the terraced rock, while the salt spray is flung high and the backwash trickles down.

OVER NIAGARA'S DIZZY BRINK rush twelve million cubic feet of water every minute; yet even figures like these do not convey a clear impression of these stupendous falls. The rush of green water in the foreground is the American Falls, which are divided by Goat Island from the foaming spray of the much wider Horseshoe Falls on the Canadian side.

MIRROR LAKE (LAKE PLACID), SUMMER VACATIONERS' PARADISE

Upper New York State has a strong lure for vacationers and sports' lovers the whole year round. Mirror Lake (above) is one of the most picturesque spots in the Adirondack Mountains.

BOTH PHOTOS, CHARLES PHELPS CUSHING

WINTER SNOWS FIND SKI ENTHUSIASTS FLOCKING TO LAKE PLACID

The long winter season makes Lake Placid a popular spot with skiers and skaters. It is also the answer to a photographer's prayer, for no season of the year is without its rare beauty.

PHILIP GENDREAU

MILKING TIME and the cows wend their way through a stony pasture in New Hampshire. Cattle can forage on such boulder-strewn soil but it is poor for the raising of any crops.

JANION FROM CUSHING

THE BEAR AND THE SEALS, amusing sculpture in ice. Such displays are a feature of the merry winter carnival held at Dartmouth College, in Hanover, New Hampshire, every year.

COPYRIGHT DETROIT PHOTOGRAPHIC CO.

DELAWARE WATER GAP is the entrance to the Pocono Mountains of Pennsylvania and the beautiful country in the extreme north of New Jersey. The river cuts its way through the Kittatinny Mountains, and divides the two states. As we look south from Winona Cliff outside the town of Delaware Water Gap, we see the Pennsylvania side (Mt. Minsi) on the right, and the Jersey side (Mt. Tammany) on the left. The famous Pocono country has few lakes, but many streams and waterfalls, and the woods are full of birches, pines and rhododendrons.

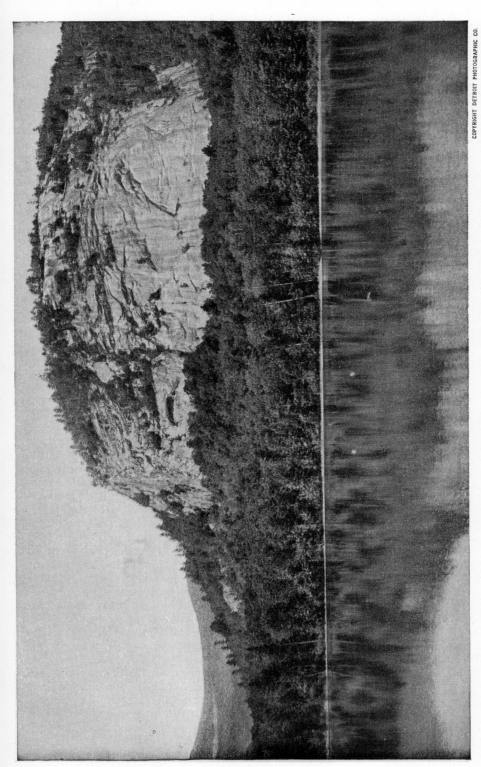

WHITE HORSE LEDGE stands out boldly above the smooth waters of Echo Lake, and the scarred cliff looks as peaceful as its reflection. Along the shore an occasional birch trunk gleams whitely through the foliage and the green of the leaves is accentuated by the darker color of the pines and hemlocks. It is remarkable how the trees can take root in tiny pockets of soil and get a foothold on what looks like bare rock. This Echo Lake is near North Conway, New Hampshire, within the White Mountains and only a few miles from the Maine border.

FINE MARBLE is cut and lifted in huge blocks from Vermont's world-famous quarries.

VERMONT MARBLE CO.

region of northeastern Pennsylvania. Riding up from Philadelphia into the Blue Ridge one comes suddenly to the coal country, with no warning except that the water in the tumbling mountain streams has become black. Green mountains give place to hills of dull black coal dirt, refuse from the "breakers" where the coal is sorted and graded after being hauled from the deep mine shafts. Railroad tracks cover every valley-floor and hug the banks of streams, and loaded freight cars by the mile stand on the sidings, waiting to be made up into trains. Where mining towns were once very dirty and unattractive, much progress has been made toward better living conditions.

Scranton, Pennsylvania's third largest city, and Wilkes-Barre have grown up in the heart of the anthracite mining area. But there is a great deal of industry other than coal in this region. Allentown, Bethlehem and Easton join Scranton and Wilkes-Barre as centers of machinery, truck, textiles, clothing and tobacco manufacture and railroad operations. Noted for its steel industry, Bethlehem is also remarkable for its many cultural interests as a Moravian community.

As abruptly as it begins, the region of coal veins ends, and the hills are beautiful once more. The Susquehanna winds its shallow way down through the mountainous centre of Pennsylvania, and is joined not far above Harrisburg by the lovely Juniata. The Great Valley of the Appalachians is well marked in Pennsylvania, and here as in other states to the south it is remarkable for its fertility and beauty. West of the mountains begins another great industrial section, and soft coal, oil and natural gas have all contributed to the growth of Pittsburgh, Johnstown, Altoona, Connellsville, Erie and other places where manufacturing is supreme, be the product glass, steel, pig-iron, coke or silk.

The westernmost tip of Maryland belongs in the soft coal country, and it is connected with the eastern part of the state by a narrow strip of land where the Great Valley swings south from Pennsylvania. Eastern Maryland and Delaware belong together geographically, for they

THE PRINCETON TIGER TAKES HIS EASE

Two huge statues of the tiger, a symbol dear to the hearts of Princeton men, guard the entrance to historic Nassau Hall on the Princeton campus. This beautiful old building, completed in 1756 and the first to be built at Princeton, suffered great damage during the Revolution. Within its walls the Continental Congress sat for several months during 1783.

are in a very real sense the product of two rivers, the Susquehanna and the Delaware. The peninsula between Chesapeake Bay on the one side and Delaware Bay and the Atlantic on the other includes all of the state of Delaware and that part of Maryland called the Eastern Shore. The western shore of the Chesapeake is also lowland built up by river silt. This level land is like New Jersey in its fertility for truck crops. Eastern Shore strawberries and Delaware peaches are known for their excellence, and they go far afield. The canning of fruits and vegetables is an important industry in Baltimore and in the entire region. The shores of Chesapeake Bay are so indented with small streams that boats can quickly and easily bring the fresh fruits and vegetables across from almost any part of the peninsula. Both the Chesapeake and the Delaware are full of fish and shellfish; the oyster beds are especially valuable, and every spring "Delaware shad" is a phrase full of pleasant meaning in near-by markets.

The factories and shipyards around Wilmington make upper Delaware look very different from lower Delaware and Maryland, where the atmosphere of the South begins to make itself felt. The Mason and Dixon line between Pennsylvania and Maryland was long the boundary between free and slave states, and south of it the Negro population is noticeable outside the cities, which is not the case in the North. Down on the Eastern Shore one can still see fine old Southern houses that once belonged to the owners of large plantations. Thus these two states are on the borderland, the transition ground where the customs and economic organization of the North began to give way, much as the climate changes to a warmer type as we go south.

The Northeast takes much of its character from the fact that it has been industrialized longer than any other part of the country. The dense population must be fed largely on imported food, and many manufacturing industries are also dependent on raw materials brought from a dis-

THE CHEW HOUSE was in the thick of the Battle of Germantown and saw heavy fighting when the Americans tried to dislodge a detachment of British troops which had occupied it. Germantown was settled by German colonists in 1683, only a year after the founding of Philadelphia, and in 1777 it was a substantial village where many prominent men lived. None of its old mansions is more stately than Cliveden, the residence of Chief Justice Chew. To-day it is well within the city limits, and street cars clatter past Cliveden's gate.

THE JOHNSTON GATE is at the west side of Harvard Yard in the heart of the busy university city of Cambridge, Massachusetts. Within the Yard the atmosphere is one of dignity and quiet, and the noise of clanging traffic seems far away. Harvard Hall, behind the gate, is one of the charming old Georgian buildings which recall the days when Harvard was small; the University has long since spread beyond the limits of the Yard. Radcliffe College is nearby and the Massachusetts Institute of Technology is also located in Cambridge.

FROM MAINE TO MARYLAND

tance. This indicates the importance of the equipment, the supply of skilled labor, and the financial resources which the region has at its disposal. From the industrial viewpoint—and from the historical also, to some extent—the Northeast is the oldest and best-established part of the country. This has varied results. New York City is the centre of the publishing business, the centre of art interest and of things theatrical, though its leadership in the last respect is less absolute than it once was. Consider also the astonishing number of first-rank educational institutions located in these eleven states: Harvard, Yale, Dartmouth, Smith, Massachusetts Institute of Technology, Mount Holyoke, Wellesley, Brown, Columbia, Vassar, Cornell, Rochester, Princeton, Rutgers, Pennsylvania, Bryn Mawr, Johns Hopkins, and dozens of others, including many small colleges of the highest grade—an almost endless list. Some of the institutions we have named have first-class professional schools, and there are others, some without university connections. Large provision is also made in special institutions for instruction and practice in the fine arts. The museums in the larger cities of the section to a great extent serve the same ends. The intellectual and æsthetic pleasures of the Northeast are among its finest features, as attractive as the beauty of stream and shore, mountain and sea, inland lake and tidewater bay.

NORTHEASTERN STATES: FACTS AND FIGURES

STATE	TOTAL AREA (SQ. MILES)	POPULATION (1950)
Maine	33,215	913,774
New Hampshire	9,304	533,242
Vermont	9,609	377,747
Massachusetts	8,257	4,690,514
Rhode Island	1,214	791,896
Connecticut	5,009	2,007,280
New York	49,576	14,830,192
New Jersey	7,836	4,835,329
Pennsylvania	45,333	10,498,012
Delaware	2,057	318,085
Maryland	10,577	2,343,001

PRODUCTION AND INDUSTRY

Although much of the land in the Northeastern states is not suited to agriculture, intensive farming has made some regions highly productive. Truck-farming and fruit-growing are important in New York, New Jersey, Pennsylvania, Delaware and Maryland; New York and Maine are among the leaders in the production of potatoes. Pennsylvania and New York have large interests in stock-farming and dairying. About one-fourth of the fish products in the United States (by value) comes from the Northeastern states. All the hard coal and about one-fifth of the soft coal of the nation are mined in Pennsylvania, which was second in the value of its mineral products until California surpassed it in 1949. The value of building stones, including granite, marble, sandstone and limestone, quarried in the Northeastern states is more than two-fifths of the nation's total. The value of manufactures in the northeast is two-fifths of the national total. Leading manufactured products are clothing, electrical and non-electrical machinery, textiles, food, iron and steel, chemicals, fabricated metals and transportation equipment. Printing and publishing are carried on. Principal ports: New York, Philadelphia, Boston, Baltimore, Providence. Shore and mountain vacation resorts are important.

POPULATIONS OF CHIEF CITIES

State capitals (1950 census): Augusta, Maine, 20,913; Concord, New Hampshire, 27,988; Montpelier, Vermont, 8,599; Boston, Massachusetts, 801,444; Providence, Rhode Island, 248,674; Hartford, Connecticut, 177,397; Albany, New York, 134,995; Trenton, New Jersey, 128,009; Harrisburg, Pennsylvania, 89,544; Dover, Delaware, 6,223; Annapolis, Maryland, 10,047.

Other important cities (1950 census): New York, New York, 7,891,957; Philadelphia, Pennsylvania, 2,071,605; Baltimore, Maryland, 949,708; Pittsburgh, Pennsylvania, 676,806; Buffalo, New York, 580,132; Newark, New Jersey, 438,776; Rochester, New York, 332,488; Jersey City, New Jersey, 299,017; Syracuse, New York, 220,583; Worcester, Massachusetts, 203,486; New Haven, Connecticut, 164,443; Springfield, Massachusetts, 162,399; Bridgeport, Connecticut, 158,709; Yonkers, New York, 152,798; Paterson, New Jersey, 139,336; Erie, Pennsylvania, 130,803; Scranton, Pennsylvania, 125,536; Camden, New Jersey, 124,555; Cambridge, Massachusetts, 120,740; Elizabeth, New Jersey, 112,817; Fall River, Massachusetts, 111,963; Wilmington, Delaware, 110,356; Reading, Pennsylvania, 109,320; New Bedford, Massachusetts, 109,189; Allentown, Pennsylvania, 106,756; Waterbury, Connecticut, 104,477; Somerville, Massachusetts, 102,351; Utica, New York, 101,531; Lynn, Massachusetts, 99,738; Lowell, Massachusetts, 97,249; Schenectady, New York, 91,785; Niagara Falls, New York, 90,872; Quincy, Massachusetts, 83,835; Manchester, New Hampshire, 82,732.

FROM VIRGINIA TO TEXAS

The Story of the Southern States

The term Southern states is sometimes applied to all the fifteen states which permitted slavery in 1860, and sometimes is restricted to those which seceded from the Union. Our list is different from either. We have included all the states which seceded, the newer states of West Virginia and Oklahoma, and also Kentucky, fourteen in all. We have treated elsewhere Maryland, Delaware and Missouri, though they were slave-holding states. It is a vast area with great variations of elevation, soil and climate, and can raise many different crops. Within the present century it has become important in manufactures, especially cotton and tobacco, and in the production of minerals, especially petroleum. Some of the states are among the oldest, though one, Oklahoma, is one of the latest admitted to the Union.

PERSONS speaking or writing about the South or the Southern states do not always mean the same thing. Fifteen states still recognized slavery in 1860, though only eleven of these seceded from the Union. Two states, usually classed as southern, have been admitted since 1860, and Missouri, though it recognized slavery, is usually called central. What are the Southern states?

The census applies the word South to three groups: South Atlantic, East South Central and West South Central. We have already mentioned two of the South Atlantic states, Maryland and Delaware, among the Northeastern states, but shall call all the others Southern. Our list then will be Virginia, West Virginia, North Carolina, South Carolina, Georgia and Florida of the South Atlantic states; Kentucky, Tennessee, Alabama and Mississippi, the East South Central; Arkansas, Louisiana, Oklahoma and Texas, the West South Central, fourteen states in all.

The oldest permanent settlements in the United States were made within this area, Spanish St. Augustine and English Jamestown; four of these states, Virginia, the Carolinas, and Georgia were of the Original Thirteen. On the other hand, the territory from which the states west of the Mississippi were made was gained from France and Mexico in the nineteenth century; Florida, East and West, was obtained from Spain; Oklahoma was Indian country until comparatively recently; Texas, after nine years as an independent republic, was annexed to the Union, but much of it was raw frontier long afterward.

The South is a vast area with many differences in elevation, soil and climate. Texas alone is much larger than all the states we have classed as Northeastern. It is farther from Richmond to Memphis in an adjoining state than from Richmond to Portland, Maine. The distance from Richmond to El Paso is longer than the farthest point in North Dakota. The

GENERAL LEE IN STONE

A pageant of the Confederacy was begun, but never completed, on the granite face of Stone Mountain near Atlanta. Here the sculptor is carving a gigantic head of General Lee.

COPYRIGHT DETROIT PHOTOGRAPHIC CO.

THE SWANNANOA and the French Broad, two clear mountain rivers, come together almost within the city limits of Asheville, North Carolina, the chief city in the famous "Land of the Sky." In this favored region Nature has been lavish of her gifts of beauty and it is difficult to decide in which season of the year the views are the loveliest. Some would vote for the early summer when the rhododendrons bloom while others would give their verdict for the autumn with its amazing blaze of color. The other seasons also have their strong advocates.

LOOKOUT MOUNTAIN rises above the Tennessee River near the busy city of Chattanooga, Tennessee. On this mountain was fought the "Battle above the Clouds," a part of a general engagement around Chattanooga during the Civil War. The Tennessee is formed by the junction of the Holston and the French Broad (page 226); it flows southwest into Alabama, thence northward across Tennessee; and finally empties into the Ohio, of which it is the largest tributary. The river is navigable for its whole length during a part of the year.

total area of these fourteen states is 887,048 square miles; and in them live (1950 census figures) 43,733,824 people —white, black and Indian—who differ widely in their attitude toward many questions.

Though there has been comparatively little foreign immigration except from Mexico, the original settlers of the older states were of many nationalities, English, Irish (North and South), Scotch, Welsh, German, French Huguenot and Swiss. The acquisition of the territory beyond the Mississippi brought in many of French and Spanish blood. Some foreigners have come in since the Revolution and there has been considerable migration from other sections, especially into Texas and Oklahoma. Speaking broadly, however, the composition of the population has been little changed.

Most of the physiographic divisions mentioned in an earlier chapter are represented in these states. The Atlantic-Gulf Coastal Plain extends the whole length of this area, from the Potomac to the Rio Grande, and along the Mississippi stretches northward to the southern tip of Illinois. Florida, Mississippi and Louisiana are altogether within this division. Parts of Virginia, North Carolina, South Carolina, Georgia and Alabama are included in the Piedmont Belt or Plateau. The Appalachian Mountains are west of the Piedmont in all these states and occur also in West Virginia, Kentucky and Tennessee. West Virginia, however, is almost entirely in the Appalachian Plateau, which also covers parts of Kentucky, Tennessee and Alabama. East of the Mississippi the Central Plains dip down through Kentucky and Tennessee into the northern part of Alabama, and to the west of that stream extend through Oklahoma into

WHEELER DAM HELPS TO HARNESS THE POWER OF A GREAT RIVER
This vast retaining wall built across the Tennessee River in Alabama by the Tennessee Valley Authority controls floods, supplies electrical power and aids deep-water navigation.

VIRGINIA STATE CHAMBER

TOBACCO LEAVES FLAP IN THE WIND ON A VIRGINIA FARM

When the tobacco leaves change from deep green to yellow, they are ready for harvesting. Usually each leaf is cut from the plant by hand, and then strung with others to be dried.

SELECTING SPECIAL LEAVES FOR VARIOUS BLENDS OF TOBACCO

After the tobacco has been cured it is taken to the factory. Here dried leaves of various kinds are sorted together in specific blends. Later they will be shredded for cigarettes.

COURTESY BALTIMORE AND OHIO RAILROAD COMPANY

HARPER'S FERRY, so called from an early settler, Robert Harper, who ran a ferry here before the days of bridges, is famous both for its situation and its history. The town itself is now in West Virginia at the point where the Shenandoah River, flowing northward, and the Potomac River come together. On the left of the picture the hills are on the Virginia side and on the right, Maryland. The United States arsenal and armory in the town were seized by John Brown in 1859, and during the Civil War the town changed hands a half dozen times.

COURTESY THE SAN ANTONIO ART LEAGUE

TEXAS PLAINS have been pictured here by Audley Dean Nicols at their most attractive time—late spring. They are bare and brown during the long dry summer while the vegetation in the eastern part of the country is still fresh and green. But the spring early and late is a gracious season; delicate grasses and flowers grow abundantly; the air is sweet with the fragrance of the *huisache* and *agarita*; every thorny shrub is in leaf. Even the scrubbiest cactus patch boasts amazing magenta and yellow flowers and stems of brave new green.

A DOG and a gun, cherished possessions of an elderly mountaineer in the Great Smokies.

On the other hand, in the higher Appalachians the vegetation is Alpine, and the climate is delightful. As a matter of fact, summer temperatures in most of the South are seldom so high as they are in Central United States, but the summers are longer, and there is little escape from the heat except along the seacoast and in the Appalachian Highlands.

Many metals and other minerals are found in the section, but only a few can be worked profitably. These few are so profitable, however, that three Southern states are found among the first seven in mineral production. During the seventeenth and eighteenth centuries, iron was smelted in the older states, but the discovery of richer deposits elsewhere closed the rude furnaces. Until the discovery of gold in California, North Carolina and Georgia were a source of gold, but today few of these mines are worked. There is some iron and much coal through the Appalachian region, and Alabama has be-

Texas. The Great Plains also extend southward into the two states last named. The Ozark-Ouachita Uplands include parts of Arkansas and Oklahoma, while in the extreme west of Texas, the Big Bend country, are the Trans-Pecos Highlands, a semi-arid region of mountains and filled valleys. Other divisions might be made. Texas, for example, is a sort of transition region. Some would separate the Edwards Plateau from the Great Plains, and the Central Basin from the Central Plains, and also call attention to the Central Mineral region, the oldest part of the state, a region of worn-down mountains, rich in minerals.

The section has the highest mountains east of the Rockies, and much low alluvial land little raised above the sea; high plateaus and fertile plains; land where much rain falls, and also land where there is too little for agriculture without irrigation. Southern Florida reaches almost to the tropics, and southern Texas nearly as far.

SHIP ISLAND light is a friendly beacon to sailors along the Gulf Coast of Mississippi.

THE SHRIMP FLEET has produced a major industry along the Gulf Coast. Vast quantities of the delicious shellfish are netted by the trawlers and shipped fresh or canned.

NEW ORLEANS has not lost the imprint of its first settlers, though those of French blood have long since been outnumbered by those of other nationalities. In the Vieux Carré, or French Quarter, many of the houses are built around a court, and these old gardens, even when they have fallen into dilapidation and decay, retain their distinctive charm.

CHRIST CHURCH, ALEXANDRIA, VIRGINIA

VIRGINIA CHURCHES built in colonial days still stand in various sections of the state. Washington was a vestryman of Christ Church, and Robert E. Lee later had a pew. Bruton Parish Church, below, is in Williamsburg. This town has been restored to the general appearance it bore in the spacious days when it was the capital and chief city of Virginia.

come an important coal, iron and steel state. West Virginia and Kentucky also rank high in mineral production because of coal, natural gas and petroleum, and Arkansas has a respectable position due to the same products. Virginia and Tennessee produce coal, Texas and Louisiana furnish much of the world's supply of sulphur, zinc is mined in Oklahoma, and Tennessee and North Carolina produce some copper.

Petroleum and Natural Gas

It is petroleum and natural gas, however, which give the section its high place in mineral values. Texas and Oklahoma lead all other states in the combined values of these two products, and Louisiana also has a considerable production, in addition to the states named in the preceding paragraph. This has been almost entirely a development during the present century.

Many of the states have building stones of good quality. Tennessee and Georgia produce much marble, and North Carolina ranks third in the production of granite. There is much stone suitable for making cement, and Florida is first in phosphate rock. The clay deposits are widely spread, and are considerably worked. Dozens of other minerals are found, and some are worked on a small scale, but these are the most important.

The South and Agriculture

There are millions of acres of land, level enough for agriculture, and the long growing season is favorable to crops. Somewhere or other in the section nearly every crop will grow. The region is the world's great source of cotton, which is the nation's largest single export. Winter wheat, corn, oats, sorghum, tobacco, clover, alfalfa, all the legumes; all the vegetables; nut trees of several sorts; fruits of the temperate zone as apples, peaches, pears and plums and cherries; sub-tropical fruits as oranges, lemons, grapefruit and figs; the small fruits as strawberries and blackberries—all of these are grown somewhere in the South.

Not all of them are grown in sections where they can easily be grown. In the cotton belt, and in the regions where tobacco flourishes, many of the farmers devote all their attention to one or both of these crops and buy most of their food, just as many of the wheat farmers of the Northwest raise wheat exclusively. If their "money crops" bring a good price they can pay their debts and are satisfied; if the crop is poor or the price is low, they sometimes go hungry. Many farmers in regions where the pasture is fair, or even good, keep few or no cows.

Better methods have been introduced into many parts of the South, however. Through the Extension Service of the Department of Agriculture, the farmers are taught how to make the best use of their soil. The Tennessee Valley Authority has also brought prosperity to the farmers of the region it serves.

Early Fruits and Vegetables

The climate gives the section great advantages with certain crops. Early vegetables from the lower Rio Grande or from Florida appear in the markets of northern cities before the snow has gone from the streets. Additional supplies from Georgia, the Carolinas and Virginia follow when seeds in suburban gardens in New England have hardly sprouted. Peaches from Georgia come to market early, to be followed a little later by others from North Carolina or Virginia. Local passenger trains must take the sidings in order that solid trains of refrigerator cars may pass on their way to the northern cities.

The South can raise more cotton than the world is able to buy, in spite of the ravages of the boll weevil. Texas is always first, Mississippi second, with Georgia, Alabama and Arkansas contending for the third place. Some cotton is grown in all Southern states except one; North Carolina, Kentucky and Tennessee grow two-thirds of the tobacco of the United States. Nearly all the sweet potatoes are grown in the South with North Carolina and Georgia leading, and peanuts are a southern crop almost exclusively.

At one time, a century or more ago, the older states of the South manufactured in small shops or little factories most of the

FORT SUMTER guarded the harbor of Charleston, South Carolina, at the time of the Secession. A Confederate attack on Union troops in the fort marked the beginning of the Civil War.

PEACEFUL AND PRIMITIVE is life in the bayou country along the Gulf Coast. The marshes and sluggish streams teem with birds and small mammals as well as water plants.

BAY SAINT LOUIS on Mississippi Sound was one of the earliest settlements in Mississippi. Attempts to colonize this section were made soon after 1715. The beauty of the town and the excellence of the white sand beach make it a favorite summer resort; recently it has become the centre for the economic life of the surrounding country, furnishing a market for fish, oysters, fruits and vegetables. The trestle which is shown at the left is a part of the Louisville and Nashville Railroad which crosses Saint Louis Bay.

PHOTOGRAPH FROM LANNEAU

MAGNOLIA GARDENS are the most widely known of the delightful gardens near Charleston, South Carolina, some of which have been famous for their flowering trees and shrubs since Revolutionary days. Each year many tourists visit the Magnolia Gardens in early spring when the azaleas flower in great banks of rich, unbroken color and the fine old wistarias bloom; and the winding gateways and moss-hung oaks make this a place of quiet beauty at any season. These gardens are on the Ashley River, twelve miles from Charleston.

AROUND THE FIRST TURN AT CHURCHILL DOWNS ON DERBY DAY
On the outskirts of Louisville, Kentucky, is Churchill Downs, the site, since 1875, of the most famous, most colorful spectacle in American horse racing—the Kentucky Derby.

OUTSIDE LEXINGTON IN THE BLUEGRASS OF CENTRAL KENTUCKY
Like their great father, Man o' War, Thoroughbred colts of Faraway Farm near Lexington, Kentucky, nibble the rich bluegrass that builds the sinews and stamina of champion runners.

Photograph by Edgar Orr

DOGWOOD IN BLOOM, DRUID HILLS, ATLANTA, GEORGIA

The flowering dogwood so common in the Southern states blooms in early spring before the leaves appear on the trees and the woods in many places are white with the broad blossoms. The tree seldom grows to be very tall. The purple Judas tree and the pink azalea may often be seen in the same woods. Druid Hills is an attractive residential suburb of Atlanta, and the streets and drives are exceedingly attractive. There are many other suburbs of this growing city which are no less pleasing to the eye, as the residents take much pride in their surroundings.

articles the people needed. As the world's demand for cotton and tobacco increased, more and more attention was given to agriculture, and manufacturing declined, though it was never entirely given up. There were numerous little cotton factories in 1860, many sugar mills, and some factories of other sorts. During the last quarter of the nineteenth century manufactures began to grow. Now there are five states, Texas, North Carolina, Maryland, Georgia and Virginia, each of which reports annually products valued at more than a billion dollars, and the production of several other states is worth noting.

The Growing Cotton Industry

The first important industry to develop was the manufacture of cotton. In 1890 New England mills used three times as much cotton as the Southern, but by 1905 the Southern mills had forged ahead, and now they use almost four times as much. Though Massachusetts still has many spindles, three states, South Carolina, North Carolina and Georgia, sometimes run more spindle hours in a year. This term means that the number of spindles is multiplied by the number of hours they run. Southern mills usually run more hours in a week and most of them have run steadily, while, for several years, many New England mills did not run full time. New England still manufactures a large part of the finer goods, but Southern mills are doing more and more of this class of work, and their total product is worth much more. North Carolina leads in the number of spindles, with South Carolina not far behind. Georgia follows with Alabama in fourth place. Tennessee and Virginia also manufacture considerable cotton. There are many knitting mills, especially in North Carolina and Tennessee. The cotton industry has not taken root in any other Southern state.

New Social Problems Appear

The growth of the textile industry introduced many new problems. The workers, nearly all white, were drawn from the rural districts, where they were tenant farmers or small landowners. Their change to factory life brought about real difficulties in adjustment. There was at first little industrial friction, as the workers appeared to be fairly contented with their wages and working conditions. Discontent eventually raised its head, however. By the year 1951, all-out campaigns had been undertaken for the purpose of organizing the textile workers of the South. It was estimated that between 15 and 20 per cent of the men and women in the textile industry had been unionized.

The manufacture of cotton is not the only industrial activity. North Carolina now manufactures more tobacco and snuff than any other state, and makes nearly half of all the cigarettes. Virginia makes another third. Florida and Virginia also make cigars. Largely as a result of the tobacco industry North Carolina pays more taxes to the Federal Government in some years than any other except New York. In the memory of many men yet living cottonseeds were thrown away. Now the products are worth over $400,000,000 a year. The oil is used for food, and in making soap and candles. Formerly some of it was exported to Europe, there mixed with olive oil, and then crossed the ocean again to be sold as pure olive oil. The meal and hulls are valuable for feeding cattle, and some of the meal is also used in making fertilizer. The refining of petroleum has become an enormous industry in the oil-producing states.

The Production of Lumber

There is considerable timber left in the South. Mississippi, Louisiana, Alabama, Texas, Arkansas, Georgia and North Carolina each produces more lumber than any states except Washington, Oregon and California. Some is hardwood, but more is pine and cypress. Though most of the lumber is exported, a considerable quantity is taken a step or two farther. High Point is the southern center of the furniture industry, and there are many other factories scattered through the South. There are several large pulp mills which are increasing their production, but the industry is not yet important. Turpentine and resin, obtained from the pine, have always been Southern products. Georgia and Florida lead.

There are hundreds of other factories producing scores of articles in the South. One advantage is the recent extensive development of hydroelectric power. The South Atlantic states generate more horse power than any other group, except the three Pacific states, and Alabama and Tennessee have extensive developments. Tennessee, Alabama and North Carolina, ranking fourth, fifth and sixth, top the other Eastern states except New York. The ease with which electric power may be transmitted puts the village on an equality with the city as a site for manufacturing, and is causing the development of many towns rather than a few great cities. It will be noted that these states which have developed large amounts of power are those a considerable portion of whose surface is mountain or plateau. The Tennessee Valley Authority (TVA), a government agency, has developed hydroelectric power, better farming methods and control of floods.

Atomic Energy Plants

World War II and its subsequent tensions left their mark on the South. When the atom bomb became a possibility in 1942, one of the first three atomic plants, known as the Clinton Engineer Works, was constructed at Oak Ridge, Tennessee. The 60,000-acre government reservation on which it is built became an "atomic town," its population composed of plant workers and their families. Houses and stores are rented from the Government, and the area has its own schools, hospitals, recreational and welfare facilities. During the war, the population reached a high of 70,000 but that had dropped at the latest census to half that number.

Early in 1952 the great Savannah River hydrogen-bomb project in South Carolina began to take shape. Ellenton, a small town on the 202,000-acre tract, was moved, lock, stock and barrel, along the highway and relocated at Jackson. This was to make way for the plant installation of the Atomic Energy Commission.

In spite of the many factories and the great atomic developments, the South is still rural. No state is "industrial-minded," though a few are becoming so. The largest city in the section ranked fourteenth in 1950.

Rural Population in the South

This rural population is interesting from many standpoints. Before the Civil War—in the states east of the Mississippi and in Louisiana—the English ideal of life on a country estate was dominant, though as in England the wealthy owner of a plantation might have a city house. Even the professional man who was forced to live in a city often had a plantation to which he hoped to retire. While, of course, there were cultured and wealthy individuals in the cities, in general it may be said that the city was considered a convenience only, if not, in fact, an evil. Most of the famous Southern mansions were on the plantations rather than in the cities. Certainly there was more culture in the rural districts.

This condition no longer exists. The destruction wrought by Civil War and Reconstruction destroyed the old plantation system, and the families have generally moved to town or city.

Problems of the Rural South

The rural situation is complicated by the presence of the Negro, who must always be considered. After Reconstruction, the people who had not money enough to maintain one efficient public school system felt themselves obliged to maintain two, with the result that neither was even moderately satisfactory. Only slowly has improvement been manifest, and even now the rural and small town schools, as a whole, are less efficient than those in other sections. While more progress has been made in some states than in others, there has been real progress in all, and the amount of illiteracy both black and white has been greatly lessened. Between 1920 and 1930 white illiteracy was reduced about one-third, and Negro illit-

THE SINGING TOWER

This superb tower, with its carillon of 71 bells, was built by the famous editor, Edward W. Bok, who is buried here. It is at Iron Mountain, the highest point in Florida.

eracy almost one-fourth. By October 1947, illiteracy had reached an all-time low throughout the country.

Twelve of these fourteen states had slavery in 1860, and there is a considerable proportion of Negroes in all of them, though that proportion is decreasing. Mississippi and South Carolina had a small majority of Negroes in 1920, though before 1930 the latter state had a majority of whites. In West Virginia and Oklahoma, the proportion is small, but in several other states it is over 35 per cent. In the section as a whole the proportion is a little less than one-fourth.

When the Negroes were freed, in 1865, few could read or write. It is a measure of their advance that almost 90 per cent can do so today. There are a number of Negro colleges and technical schools; and Negro race is no bar to admission in many of the country's leading universities. By law Negroes are supposed to have educational opportunities equal to those provided for whites. This is still not true in some parts of the South, but the picture is changing by slow degrees.

Even before slavery was abolished, a few Negroes achieved fame, among them Joshua Johnston, Benjamin Banneker and Frederick Douglass. Today there are hundreds of Negroes who are outstanding, in almost every field—music, art, science, literature, sports—and many of these men and women were born and brought up in the South. Relations between the colored and white people are still not easy, but a hopeful note in the situation is that both sides today are providing responsible leadership.

Reference has been made to the tenant system, which has handicapped Southern agriculture. While, perhaps, the best system possible when it was put into effect after the Civil War, its continuance has resulted in unsatisfactory farm conditions. Though a number of tenants, both white and Negro, do manage to purchase land, many of the farms are still operated by tenant farmers and sharecroppers. Because they are often short-time occupants, they do not have the same interest as a farmer-owner in improving the land

CHARLES PHELPS CUSHING

A POWER SHOVEL loads railroad dump cars with tons of bauxite ore, near Bauxite, Arkansas. The region is the United States' chief home source of the ore, which yields aluminum.

MONKMEYER

AN OIL PUMP in the middle of the street reminds visitors to the town of Barnsdall, in northern Oklahoma, that it is in the midst of a fabulously rich petroleum-producing area.

Key West Chamber of Commerce

A BEAUTIFUL SPECIMEN OF FISHERMAN'S LUCK

From the southern tip of Florida a long succession of coral islands (called keys) extends in a southwesterly direction. From Key West, at the tip of the island chain, sportsmen go out in boats large and small, lured by tarpon and other large fish that abound in the deep waters. The tarpon pictured here is a summer fish, though occasionally one may be caught in winter.

or preserving the equipment, and so this tenancy is unsatisfactory to both the owner and the tenant and has an adverse effect on the community as a whole. Recently there has been some reconversion of plantations from sharecropper to hired labor and the general standard of farming is rising, although slowly.

One cause of the backwardness of the rural population in some states was the isolation caused by roads that were almost impassable in winter. Recent years have marked a tremendous improvement, and all the Southern states now boast highway systems that include from one to thirteen thousand miles of high-type surfaced roads, with more constantly being constructed. The hordes of motorists who head south each winter give ample proof of this change.

The schools in the towns and cities are generally good, and some invite comparison with those of any other section, in buildings, standards of instruction, and qualifications of the teachers. There is more property to tax in the towns and cities, and in some cities the citizens have voted heavy rates for school purposes.

Some of the oldest colleges and universities in the country are in the South. The College of William and Mary at Williamsburg, Virginia, was founded (1693) next after Harvard, and the University of North Carolina was the first state university to begin operation, graduating the first class in 1795. Thomas Jefferson wore with pride the title "Father of the University of Virginia." Every state has a state university, and while not all are real universities, some are recognized as first-class institutions in every respect. Private philanthropy has established other universities, some under the auspices of a church and others independent. Duke University at Durham, North Carolina, through the will of the tobacco and power magnate, James B. Duke, is one of the best endowed institutions in America with imposing buildings, admirably equipped. Rice Institute in Houston, Texas, is an-

THE WONDERFUL BEACHES ALONG THE EAST COAST OF FLORIDA

Photograph by R. H. LeSesne, Courtesy Florida East Coast Railway

Many of the beaches of East Florida at low tide are as smooth and hard as a floor. The famous Ormond-Daytona beach is twenty-five miles long and 300 feet wide, and is so hard that motor races have often been held upon it. In one race Major H. O. D. Seagrave in his Mystery S attained the speed of 203 miles an hour for the course. In recent years most of the major racing has been done on the beach around great Salt Lake in Utah. During World War II many of the large Florida resort hotels were turned into barracks to house Army students in many branches of the service.

other institution amply endowed by a millionaire which has high standards.

There are scores of denominational colleges, too many perhaps. In some cases a denomination has established several colleges in the same state. Many do excellent work, but others are handicapped by the lack of sufficient funds. Some of these have voluntarily dropped into the class of junior colleges.

The cities of the South have grown rapidly, becoming busy centers of manufacturing and trade. Oil in the southwest, shipping from ports along the Atlantic and Gulf coasts, the tourist trade in Florida and a general industrial advance throughout the region have contributed to the great increase in the number of cities with more than a hundred thousand people. In 1940 there were twenty; the 1950 census showed twenty-nine. Texas had seven; Tennessee, four; Alabama, Florida and Louisiana, three each; Virginia Georgia and Oklahoma, two each; and Arkansas, North Carolina and Kentucky each had one. Only West Virginia, South Carolina and Mississippi were without any cities of more than a hundred thousand.

The tourist industry is becoming important in the South. All through the Appalachian country there are excellent hotels, which attract visitors from the South in summer, and from the North in winter. Asheville has become famous as having some of the finest resort hotels to be found anywhere. The "sandhill country" of North and South Carolina attracts many winter visitors, and Georgia also gets its share. Florida attracts increasing numbers of winter residents, and many have established permanent homes. Miami, Palm Beach, Tampa, St. Petersburg and a dozen other towns are known everywhere. All along the Gulf of Mexico there are winter colonies, and New Orleans has always attracted many visitors.

THE SOUTHERN STATES: FACTS AND FIGURES

STATE	TOTAL AREA (SQ. MILES)	POPULATION (1950 CENSUS)
Virginia	40,815	3,318,680
West Virginia	24,181	2,005,552
Kentucky	40,395	2,944,806
Tennessee	42,246	3,291,718
North Carolina	52,712	4,061,929
South Carolina	31,055	2,117,027
Georgia	58,876	3,444,578
Florida	58,560	2,771,305
Alabama	51,609	3,061,743
Mississippi	47,716	2,178,914
Arkansas	53,102	1,909,511
Louisiana	48,523	2,683,516
Oklahoma	69,919	2,233,351
Texas	267,339	7,711,194

PRODUCTION AND INDUSTRY

Agriculture is the most important occupation in all of the Southern states except West Virginia. The extent of the area with its variety of soil, climate and rainfall makes it possible to grow many temperate and semi-tropical crops. The most important crops are cotton (these states furnish over 70 per cent of the world's supply), tobacco (North Carolina, Kentucky, Virginia, South Carolina and Tennessee furnish over 80% of U. S. supply) and sugar-cane (Louisiana). The South Atlantic and Gulf Coast states produce quantities of early vegetables. Virginia raises large quantities of apples, North Carolina and Georgia, peaches, Florida and Texas, citrus fruits. Dairying has become important in these states; one-fourth of dairy cattle of the United States are now found here; Texas raises many beef cattle and sheep. Rich forest and mineral resources. Among the mineral products are petroleum (over one-third of the world's supply comes from Texas, Oklahoma and Louisiana), iron ore and coking coal, sulphur (Louisiana and Texas), bituminous coal (West Virginia), bauxite and a variety of building materials.

IMPORTANT CITIES

Populations of state capitals (census of 1950): Richmond, Virginia, 230,310; Charleston, West Virginia, 73,501; Frankfort, Kentucky, 11,916; Nashville, Tennessee, 174,307; Raleigh, North Carolina, 65,679; Columbia, South Carolina, 86,914; Atlanta, Georgia, 331,314; Tallahassee, Florida, 27,237; Montgomery, Alabama, 106,525; Jackson, Mississippi, 98,271; Little Rock, Arkansas, 102,213; Baton Rouge, Louisiana, 125,629; Oklahoma City, Oklahoma, 243,504; Austin, Texas, 132,459.

Population of other important cities (1950): Houston, Texas, 596,163; New Orleans, Louisiana, 570,445; Dallas, Texas, 434,462; San Antonio, Texas, 408,442; Memphis, Tennessee, 396,000; Louisville, Kentucky, 369,129; Birmingham, Alabama, 326,037; Fort Worth, Texas, 278,778; Miami, Florida, 249,276; Norfolk, Virginia, 213,513; Jacksonville, Florida, 204,517; Tulsa, Oklahoma, 182,740; Charlotte, North Carolina, 134,042; Chattanooga, Tennessee, 131,041; El Paso, Texas, 130,485; Mobile, Alabama, 129,009; Shreveport, Louisiana, 127,206; Knoxville, Tennessee, 124,769; Tampa, Florida, 124,681; Savannah, Georgia, 119,638; Corpus Christi, Texas, 108,287; St. Petersburg, Florida, 96,738; Beaumont, Texas, 94,014.

STATES OF LAKE AND PLAIN

Farms, Mines and Mills of the North Central States

The twelve North Central states, commonly known as the Middle West, lie chiefly in the level or rolling prairies of the Central Plains, though they verge to the north into the wooded lake-dotted Superior Highlands and in southern Missouri into the Ozark Mountains. Here lie the great corn and wheat belts, a fine dairy region, rich mines and progressive manufacturing cities. The total area, comprising about a quarter that of the continental United States, supports nearly one-third of the population, produces more than a third of the crop values and contains ten of the twenty-five largest cities. It is a region largely of one-man farms, served by exceptional rail and water transportation facilities and of such manufactures as meat-packing, flour-milling and in particular, the making of automobiles and farm machinery. The people are hardy, industrious, keen to secure education. The climate is one of moderate rainfall, stimulating cold winters and of the hot summers so beneficial to the staple crops grown in the region.

THE territory we are classing as the North Central states—popularly known as the Middle West and latterly, sometimes, as the Midlands—includes the two groups of states called in the United States Census the East North Central and the West North Central states. These twelve states have a total area of 769,126 square miles. The East North Central states are Ohio, Indiana, Illinois, Michigan and Wisconsin; and the West North Central states are Minnesota, Iowa, Missouri, Kansas, Nebraska, South Dakota and North Dakota. The total area is about a quarter that of continental United States.

There were, in 1950, over 44,000,000 people in the Middle West, 124 to the square mile in the East North Central states and 27 in the Western division as contrasted to 50.6 per square mile for the United States as a whole and 300 for the Middle Atlantic states. The East North Central states have more than four times the density of population found in the West North Central states.

The population is overwhelmingly white, and over four-fifths native born. The remainder come chiefly from northern Europe. Missouri, Ohio and Illinois, however, have a considerable Negro population and the Dakotas and Minnesota have considerable numbers of Indians. The foreign-born (not many today) and those of immigrant parentage have mostly a Scandinavian or German background. Almost two-thirds of the population lives in cities and towns with 2,500 or more inhabitants. Approximately one-sixth of the people live on farms, and the remainder in small towns and villages. The population trend is toward urban communities.

The Central Plains cover most of the region. There is, however, a border of Superior Highlands in upper Wisconsin and Minnesota, a lovely region of woods and lakes, a bit of the Appalachian Plateau in eastern Ohio, and the Ozark Plateau covers a large part of southern Missouri. Down the centre of this group of states flows the mighty Mississippi, "Father of Waters." Its great tributaries are the busy Ohio and the muddy Missouri —which is almost as long as the larger stream. Between the Missouri and the Mississippi lies Iowa, the champion corn state, in the heart of the Corn Belt that extends from Ohio to western Nebraska in a great east-to-west ellipse. Here the smooth fertile soil left by the last ice-sheet, the five months of summer sunshine with warm nights and the occasional thunder-showers, make ideal growing conditions for that cereal native to American soil which the Indians taught the first white settlers to cultivate. As hogs require grain, these animals are the natural complement to the cornfields; and indeed much of that crop goes to feed the hogs and but one-tenth is used for human food.

THEY'RE OFF! The five-hundred-mile Memorial Day automobile race held on the Indianapolis Speedway is known all over the world. The first of the races was held in 1911.

North of the Corn Belt around the Great Lakes is a strip of country that depends heavily on hay and dairying. There are two huge wheat-growing areas, the northern devoted to spring wheat, the southern to winter wheat which is planted in the fall. Of course the farms of these several outstanding regions are not confined entirely to the produce named. A certain amount of general farming is the rule and portions of Michigan, tempered by the lakes, specialize in peaches and other fruit. Some timber still remains of the forests that once covered the northern fringes of these states and other areas, such as the Ohio Valley.

There is, finally, a region rich in minerals which, for this reason—coupled with the ease of transportation by water and by the straight-laid rails of the prairie country—has become a great manufacturing region. The lower lake region, with an accessible supply of metals, coal and wood, has indeed become the heart of the automobile industry and a center for the manufacture of farm machinery.

Before we quote figures to show the immensity of both farming and manufacturing in this rich region, let us see what kind of people have settled here. Because

MONUMENT to a Potawatomi chief on the site of an Indian village, northern Indiana.

of early exploration the French claimed the Mississippi Valley and hoped to keep the English east of the Alleghenies. After years of conflict with Indian allies on both sides, England won, though traces of French occupancy still persist. Most of the territory of the West North Central states was included in the Louisiana Purchase which came to the United States from France in 1803.

Before the Revolution had ended hardy pioneers had made their way into the region, and the migration from the East and the Southeast grew during the early years of the Republic until it became a flood. Tired of the attempt to wrest a living from the stony soil of New England, thousands made their way to the fertile lands beyond the Ohio. Later other thousands of Southerners sought greater opportunities in the West. Some loaded all their household goods into wagons and made the long trek. Others came by way of the Erie Canal after it was opened; still others reached the Ohio and floated down until they came to their destination. Later the foreign-born came —chiefly Germans and Scandinavians at first—until now these states form a cross section of the United States of today.

TOLEDO CHAMBER OF COMMERCE

THE AMPHITHEATER of the Toledo, Ohio, Zoo provides a lovely setting for a concert.

EWING GALLOWAY

THE WOOD YARD of a paper mill in Munising, Michigan, is piled high with the raw material needed for pulp and paper. Munising is on Lake Superior, in a lumbering and farming region.

THE KEOKUK DAM AT THE FOOT OF THE CANAL AROUND THE DES MOINES RAPIDS, IOWA

The Federal Government built the first lock of the Mississippi River, Gulf water. This dam, a mile and a half long, was constructed primarily with a lift of forty-one feet, at Keokuk, Iowa, which is 1,461 miles from as part of a great hydroelectric power project.

With characteristic energy the settlers of these Midlands converted them into what is today a region of fertile farms and progressive cities. Highways crisscross the countryside; today's farmhouse is equipped with such modern comforts and conveniences as telephones, radios, television, automobiles and many labor-saving devices. Some farmers even own airplanes, which they use in connection with their work as well as for pleasure. The region has its own charm of sparkling white winters, summers of goldenrod, wild roses and black-eyed Susans, autumns—in the hardwood regions—of gorgeous red and yellow foliage, of ripening grapes, and, later, hazelnuts, hickory-nuts and black walnuts. There are still deer near the Canadian border, and the early settlers used to be able to shoot prairie chickens on the plains.

The twelve states under consideration are all important agriculturally, those with extensive mines and manufactures less so than the others, however. The state with the greatest percentage of land in farms is Nebraska, and of over 47,000,-000 acres in farmland, about 19,000,000 are sown to cultivated crops, chiefly corn and oats. The state with the next greatest percentage of land in farms is Iowa. This state has almost twice as many swine as any other state, and more cattle than any state except Texas.

Half a century ago the typical farm, nearly self-sufficient with its mixed crops and home manufacture of butter, bacon, clothing and other necessities, was laboriously cultivated by human labor, chiefly that of the immediate family. Today the larger corn-grower, with his tractor and four-row planter, has been known to cover as much as forty-six acres a day, and with a four-row cultivator he can often cultivate sixty-five acres in a ten-hour working day. In the old days he could harvest only one or two acres of corn a day. Now that he has a corn-picker and husker operated by tractor, he can gather eight to ten acres a day. This complicated machinery not only appreciably reduces the need of human labor on the farms, releasing large num-

EWING GALLOWAY

SHIPS THAT PASS IN THE LOCKS OF THE SOO
A passenger ship and a freighter move in opposite directions through the Sault Sainte Marie Canals. They avoid rapids in the St. Mary's River and connect Lake Huron and Lake Superior.

BOEING AIRPLANE COMPANY

PRODUCTION LINE OF STRATOJETS IN A WICHITA, KANSAS, PLANT

The Midwestern city of Wichita has for years been a leader in the manufacture of airplanes and plane parts. Thousands are employed in its aircraft factories. Situated on the Arkansas River in the southeastern part of Kansas and reached by highways and major railroads, its manufactured items are easily shipped throughout the United States and to other countries.

bers of young people—who tend to go to the cities—but it very greatly reduces the cost of harvesting the mammoth crops after the initial investment in such machinery has been met. The figures that will shortly follow show the extraordinary quantities of grain produced.

One even hears of an occasional woman farmer, as, for instance, a large wheat farmer of western Kansas who harvests her 4,500 acres by hiring employees in three eight-hour shifts. These men operate seven combines, each of which cuts a twenty-four-foot swath, while at night they hitch the tractors to plows and turn the stubble under. Meals are served from a commissary car that follows the tractors. Of course there are many smaller farms which employ simpler machinery. The typical farm is said to be that operated by one farmer, his sons and a hired man. However, in rush seasons, plowing is often done at night.

The North Central states, taken as a whole, produce two-thirds of the corn crop, more than half of the wheat crop, over three-quarters of the oats and over three-fifths of the barley.

Where corn and wheat do not grow so well, hay and dairy animals thrive. The North Central states produce over two-thirds of the butter made in factories in the United States and about the same proportion of the butter made on farms. Wisconsin leads the country in cheese production, making more than half of all produced.

Kansas City (neighbor to Independence, Missouri, which was once the terminal of the Santa Fe cattle trail) has an important meat-packing industry. The livestock raised in Kansas includes nearly three times as many cattle as swine and a rather small number of sheep. Its stockyards and packing plants are the second largest in the world. But the products of the meat-packing firms in Illinois (chiefly in Chicago) are worth much more. Chicago's packing plants have, in fact, played an appreciable part in making that city the commercial capital of the North Central states.

Chicago is the world's largest food-distributing center. The refrigerator car made it possible for Chicago to slaughter and pack meat for the world. It has an entire section known as Packing Town with a square mile of stockyards. Milwaukee has been likened to a miniature Chicago. When Michigan lumber was plentiful, Grand Rapids and Saginaw started as sawmill towns and later became centers for the manufacture of furniture. Certain firms of the region introduced the now widely used knock-down furniture, which may be shipped flat, and knock-down garages and houses made in sections which may readily be put together. These, like the clothing and other products of the big Midwestern mail-order houses, may be sent to remote regions by rail or water. It is part of the reason why the words "standardization" and "efficiency" have come to be applied to things American.

It has been said that the lower lake region started as "a Yankee outpost of New England." This region, with its cheap water transportation, its wool and metal, has become the heart of the automobile industry. Since the turn of the century this has been perhaps the chief element in multiplying the population of Detroit by six, as well as greatly stimulating the growth of Flint, Lansing, Toledo, Cleveland and Akron. The last three are tire centers. Cleveland and other cities along the south shore of Lake Erie, where coal and iron ore meet, manufacture machinery on a large scale.

Ease of transportation, whether by water or by rail, has been an important factor in the growth of St. Louis. The nearness of the wheat to be milled or stored for shipment has helped make Minneapolis, Milwaukee, Indianapolis and Kansas City, and has formed one of many factors in the growth of Chicago—

UNION PACIFIC RAILROAD

THE SPACIOUS BUILDINGS and grounds of Boys Town, just west of Omaha, Nebraska. The famous "town" was founded by Father Flanagan to give neglected boys a better start in life.

second in size only to New York City. Chicago is a huge transportation center and has many miles of belt-line rails for shifting freight from road to road. The city is also a distribution center, famous for its mail-order houses, from which one may buy almost every possible product. Chicago is convenient to the grain regions, and its constant need for manufactured goods often keeps the freight carriers two abreast—loaded on their return trips. Thanks to the canals, the Great Lakes form one navigation unit. Six-hundred-feet freighters are so constructed that they can negotiate canals, locks and artificial channels; and besides the freighters, there are fine passenger steamers. Chutes and gigantic scoops make loading and unloading possible with incredible rapidity. Iron ore thunders, steel clangs, lumber echoes, cattle bellow, soft-coal billows; and the mingled odors of all these activities combine to give an impression as different as possible from the papery rustle of cornfields and the sunny peace of waving wheat or the lovely lake resorts that are all within the region that the metropolis serves.

The upper lake region was once high mountains but has been worn down by streams and weather almost to a plain, with hills of harder rock which contain lodes of iron and copper. Lakes are abundant in the ice-formed basins. Minnesota has 11,000 such lakes, Michigan 4,000 and Wisconsin 2,000. Unfortunately there is considerable territory covered by muskeg, swamp and rock. The forests, chiefly pine and spruce, maple and hemlock, once attracted Eastern lumbermen, and upper Michigan, Wisconsin and Minnesota successively led in lumber production, the work of which the deep snows facilitated. But so thoroughly was the timber crop reaped without replanting that tremendous problems in forest conservation were left. In many places nothing but jack pine grows. Wisconsin, however, still supplies an important paper industry.

BY STERN WHEELER THROUGH THE DELLS OF THE WISCONSIN
In south-central Wisconsin, the Wisconsin River has carved an eight-mile gorge—the scenic waterway called the Dells. The deepest part has sandstone walls, sculptured by the water.

EWING GALLOWAY

GIANT SENTINELS! GRAIN ELEVATORS AT A GREAT LAKES HARBOR
The huge, concrete cylinders at the harbor of Duluth are a familiar sight in the wheat-growing states. Often grain is cleaned, blended and dried in the elevators, as well as stored.

ARCH ROCK is an interesting natural phenomenon on Mackinac Island, Michigan, which lies in the Straits of Mackinac, connecting Lakes Huron and Michigan. The island is about two miles by three, and half the area is now included in a state park. There was a French post on the island very early, which was later overtaken by the British, and a strong fort was established. After the island fell to the United States the fort was maintained until 1895. The island is now a popular summer resort visited every year by thousands of pleasure-seekers.

FORT SNELLING still looks down protectively on the junction of the Mississippi and Minnesota rivers. Built in 1820 as an outpost in the Indian country, it was originally called Fort Anthony, but the name was changed in 1825 to honor Colonel Josiah Snelling, its first commander. Zachary Taylor, who later became president, was commander of this post from 1828 to 1829. The fort provided a base for exploration and later was a center of defense for the small settlements in the territory. The great industrial city of Minneapolis was one of these early villages.

CATERPILLAR TRACTOR CO.

MACHINERY does the work of many men in picking and husking corn on a farm near Nodaway, Iowa. Good food crops in this region mean a great deal to people all over the world.

Along the south shore of Lake Superior, where Indians were once able to secure copper at the surface, mines must now be driven very deep to get at the large deposits of the metal. In the hills around the western end of the lake, iron ore is easily mined. The great quantities of ore that this area has yielded have helped build the great steel-making centers in Ohio, Indiana, Illinois, Michigan and Minnesota, as well as those in the East. By far most of the iron ore of the United States comes from the land at the head of Lake Superior.

Kansas is one of the leading oil-producing states of the nation. Illinois is also a source of considerable petroleum in addition to being a major coal-mining state. It is worth noting that the Black Hills district makes South Dakota, instead of one of the far western states, the nation's largest producer of gold. Missouri leads the United States in mining lead, Ohio as a source of lime, and Michigan as a producer of salt and gypsum. The belt running through Ohio, Michigan, Indiana and Illinois yields much of the nation's stone, gravel, sand and clay.

Of the five principal manufacturing states, three—Ohio, Illinois and Michigan—are in the North Central region. Indiana, Wisconsin and Missouri also rank very high in this respect. Leading manufacturers include automobiles, farm machinery, household appliances and products of iron and steel, rubber and clay.

The North Central states have excellent educational facilities, including sixty thousand public elementary and secondary schools and more than five thousand private schools. This region has approximately six hundred universities, colleges, junior colleges and professional schools, a third of which are public-supported. Each state has one or more state universities. Among the largest of these are Ohio State University, the universities of Illinois, Minnesota, Michigan, Wisconsin and Indiana, and Michigan State College. Ranking with the above in the number of students are Wayne University of Detroit, operated by the city, and the University of Cincinnati. One feature of many of these institutions is that they offer extension courses.

The state universities have grown up on lands that were granted long ago by the Federal Government or were purchased with Federal funds. These schools have come to play an important part in the democratic life of the region. They not only provide a liberal college education at a low cost to thousands of deserving students but through the years they have assumed tasks not generally a part of a university's program. They carry on extension classes in all parts of their states, correspondence courses and technical research centers that often develop important new products and techniques. Staffs of experts from these centers help to keep farmers and businessmen abreast of modern advances. Apart from their university programs, the states look after many other phases of public education. Indeed, we

FUEL AND FOOD. Oil wells stand erect against the Kansas sky, altering the agricultural look in the heart of the winter wheat belt. Kansas ranks high among petroleum states.

STANDARD OIL CO. (N. J.)

PHOTOGRAPH BY EWING GALLOWAY

THE CUMBERLAND or National Road was planned to extend from tidewater over the mountains to the Mississippi River, but was never completed by the national government though several of the states attempted to continue the work. This tavern at Lafayette, Ohio, was built in 1837 by a man from Connecticut when the road through this part of Ohio was first laid out; and the building is still used as a tavern. Once the townspeople regarded the cook stove in the tavern as an invention of the devil, designed to keep men from an honest day's work.

CEDAR RIVER, or Red Cedar River, rises in Minnesota and flows nearly across eastern Iowa, emptying finally into the Iowa River, about thirty miles before it, in turn, flows into the Mississippi. The fall is considerable, as shown by the names of such towns on its banks as Cedar Rapids and Cedar Falls. Though not a large stream, the river drains a beautiful and fertile region. These bluffs, or palisades, though not to be compared in height with those of the Hudson, make the scenery along certain parts of its course distinctly attractive.

CHEESE-MAKING, ONE OF THE MAJOR INDUSTRIES OF WISCONSIN
At a Swiss cheese factory in Madison, Wisconsin, the curd is lifted from the whey to drain. The curd is the basis of cheese, and the watery, vitamin-rich whey is used for other foods.

cannot begin to name all of the opportunities for practical education in this part of the country.

Some great scholars are in the universities of the Middle West and some of the most popular poets and novelists were born in one or another of these states. The writers include George Ade, Theodore Dreiser, Mark Twain, Sinclair Lewis, Edgar Lee Masters, Booth Tarkington and James Whitcomb Riley.

The section has also produced many women of historical importance. One was Jane Addams, the celebrated humanitarian, who founded—and directed for forty-six years—Hull House in Chicago, the first and most famous settlement house in the United States. Another noted woman was Carrie Chapman Catt, the reformer. Largely through her efforts, the country adopted the Nineteenth Amendment, in 1920, giving women the vote.

Moreover, the people of these twelve states have played a great part in the national political scene. Among such activities have been the presidential nominating conventions. Between 1868 and 1952 inclusive, 18 of the 22 Republican and 18 of the 22 Democratic conventions have been held in large cities in this region.

In addition, from these states have come nine of the seventeen presidents since the Civil War—Ulysses S. Grant, Rutherford B. Hayes, James A. Garfield, Benjamin Harrison, William McKinley, William Howard Taft, Warren G. Harding, Herbert C. Hoover and Harry S. Truman.

The people are fond of calling their region by such names as the "Heart of America" or the "Valley of Democracy." They have some justification for their boasts. They are, on the whole, democratic; the level of intelligence is certainly up to, if not above the average, and the same may be said for the standard of morals. The average of material comfort is rather high, and the prospects for future development are bright.

STATES OF LAKE AND PLAIN

THE NORTH CENTRAL STATES: FACTS AND FIGURES

STATE	TOTAL AREA (SQ. MILES)	POPULATION (1950 CENSUS)
Ohio	41,222	7,946,627
Indiana	36,291	3,934,224
Illinois	56,400	8,712,176
Michigan	58,216	6,371,766
Wisconsin	56,154	3,434,575
Minnesota	84,068	2,982,483
Iowa	56,280	2,621,073
Missouri	69,674	3,954,653
North Dakota	70,665	619,636
South Dakota	77,047	652,740
Nebraska	77,237	1,325,510
Kansas	82,276	1,905,299

COMMERCE AND INDUSTRIES

The North Central states produce nearly one-half of all the farm crops in the United States, over half of the wheat crop, three-fourths of the corn crop, four-fifths of the oat crop and more than two-fifths of the barley crop. Enormous quantities of hay, potatoes, rye, buckwheat, sugar-beets, tobacco and fruits are grown each year. Throughout these states there are active livestock industries; dairying is highly developed; hogs, cattle and sheep are raised for the market. More than four-fifths of the iron ore of the United States comes from the Lake Superior "ore lands" in Minnesota, Michigan and Wisconsin. Illinois ranks high among the states in the production of coal; it is also mined in Ohio, Indiana, Kansas and Missouri. Michigan leads in the production of salt and is important for its copper. Missouri mines large quantities of lead, and the quarries throughout the North Central states yield sandstone, limestone and gypsum; there is an enormous output of Portland cement, bricks and tiles and other clay products. Silver and gold are mined in South Dakota. These states produce, by value, one-third of the manufactured products of the United States; three-fourths of the farming implements and over one-half of the world's output of automobiles are manufactured in Michigan. Furniture and paper products are important around the Great Lakes. The cereal crops and livestock industries have given rise to numerous flour and meat-packing plants.

POPULATIONS OF CHIEF CITIES

State capitals (1950): Columbus, Ohio, 375,901; Indianapolis, Indiana, 427,173; Springfield, Illinois, 81,628; Lansing, Michigan, 92,129; Madison, Wisconsin, 96,056; St. Paul, Minnesota, 311,349; Des Moines, Iowa, 177,965; Jefferson City, Missouri, 25,099; Bismarck, North Dakota, 18,640; Pierre, South Dakota, 5,715; Lincoln, Nebraska, 98,884; Topeka, Kansas, 78,791.

Other important cities (1950): Chicago, Illinois, 3,620,962; Detroit, Michigan, 1,849,568; Cleveland, Ohio, 914,808; St. Louis, Missouri, 856,796; Milwaukee, Wisconsin, 637,392; Minneapolis, Minnesota, 521,718; Cincinnati, Ohio, 503,998; Kansas City, Missouri, 456,622; Toledo, Ohio, 303,616; Akron, Ohio, 274,605; Omaha, Nebraska, 251,117; Dayton, Ohio, 243,872; Grand Rapids, Michigan, 176,515; Youngstown, Ohio, 168,330; Wichita, Kansas, 168,279; Flint, Michigan, 163,143; Gary, Indiana, 133,911; Fort Wayne, Indiana, 133,607; Kansas City, Kansas, 129,583; Evansville, Indiana, 128,636; Canton, Ohio, 116,912; South Bend, Indiana, 115,911; Peoria, Illinois, 111,856; Duluth, Minnesota, 104,511.

EWING GALLOWAY

STEEL PLANT IN YOUNGSTOWN, OHIO, DURING THE NIGHT SHIFT
Ohio is the geographical center of steelmaking in the United States. Its leading industry is iron and steel, and most of its other industries, in general, are based on these materials.

THE WHITE RIVER, which flows through the Ozark Plateau, rises in northwest Arkansas and makes its way into southern Missouri, where it winds through rock terraces. In olden times, the Shawnee and Delaware Indians established villages along its banks when they were forced to move before the advance of the white man. A common means of transportation across Ozark streams, such as the White River, is a primitive ferry. The driver of the flatboat propels his craft over the river by pulling upon a rope or wire that has been stretched across the stream.

THE BLACK HILLS of South Dakota, extending into Wyoming, are a mountainous region covering about 6,000 square miles. They were named because about one-third of the area was covered with dark pines when first visited by white men. The region is rich in minerals, particularly gold. The climate is pleasant, rainfall is abundant, and there is considerable fertile land which affords excellent pasture. Much of the area has been set aside for state and national parks and game reserves. Our picture shows a female elk within a reservation.

PRAIRIE SENTINELS—grain elevators at Hatton, North Dakota. Wheat from the valley of the Red River of the North is stored here, awaiting shipment to markets at home and abroad.

PHOTOS, STANDARD OIL CO. (N. J.)

THE CONSOLIDATED SCHOOL at Sterling, North Dakota, takes the place of many little schoolhouses once scattered over the countryside. Busses bring children from a distance.

THE EERIE LANDSCAPE of the Badlands of South Dakota. Wind and weather have carved the dry plateau into strange pinnacles and rifts, where scarcely a blade of grass can grow.

PHOTOS, STANDARD OIL CO. (N. J.)

TRIBUTE TO KING CORN—the Corn Palace in Mitchell, South Dakota. All of the decorations, including the murals, on the fantastic building are made of corn, both kernels and cobs.

WISCONSIN, in part, escaped the ice-sheets which long covered so much of the surface of the North Central states. In fact there is a whole section of considerable size known as the "driftless area", over which an ice sheet never passed and which therefore is particularly interesting from a geological standpoint. It presents numerous outlying hills which show curious sandstone formations like Castle Rocks near Camp Douglas. Such a formation is the result of erosion, and to-day is more commonly found in the semi-arid regions farther west.

PHOTOGRAPH BY EWING GALLOWAY

STARVED ROCK is a steep rocky hillock on the southern bank of the Illinois River almost midway between the cities of Ottawa and LaSalle, Illinois. Here the French explorers LaSalle and Tonty, in 1682, established Fort St. Louis which was occupied until about 1718. About 1770, some Illinois Indians were closely besieged on the rock by their enemies, the Potowatomi, and were finally starved to death. The area which includes the Rock is now a state park. Our picture shows Horseshoe Fall on what is now only a small stream in the park.

WINDOW ARCH, eroded through a wall of rock, offers a striking view of the Canyon de Chelly. The arid Colorado Plateau of northern Arizona is gashed by many such river canyons

The States Toward the Sunset
The Mountain and Pacific Coast States

What we call the West includes the eight Mountain states, in which there is an average of 5.9 persons to the square mile, and the three Pacific Coast states, in which the average is 45.3 persons as contrasted with 300 for the Middle Atlantic states. This territory, romantic since the days of the covered wagons and the gold rush of Forty-Nine, includes over a million square miles, about two-fifths of the area of the continental United States. It contains the highest mountains, the driest deserts and (around Puget Sound) the wettest mountain slopes of the country, as well as the most national parks and forests. The West was settled by those with the courage to face a remote wilderness, and there has been swift progress—skyscrapers for the larger cities, paved motor highways, airports, machine-logging, irrigating, mining, sheep and cattle-rearing, all on a stupendous scale. The ports of the Pacific coast—Los Angeles, San Francisco, Portland and Seattle—are among the busiest in the nation.

THE area we have classed as the West includes two groups of states as given in the United States Census—the Mountain and the Pacific states. It is the largest of our four divisions, including as it does eleven states with a total area of 1,187,753 square miles. The Mountain states are Montana, Idaho, Wyoming, Colorado, New Mexico, Arizona, Utah and Nevada; the Pacific states are California, Oregon and Washington. Generally they are large compared with the states of the East, and the population of most of them is still sparse, though the Pacific group is becoming somewhat thickly settled.

Most of this region was once beneath a great inland sea. Marine fossils have been found eleven thousand feet high in the various ranges of the Rockies, and California has what are called raised beaches which are as much as fifteen hundred feet above the present sea level. For the most part the mountains have undergone a slow elevation through long ages. In the arid southwest there are vast tracts where the sandstone has been eroded into buttes and canyons. Along the Pacific, a line of volcanoes once flamed red where now rise the mountain ramparts. These, too, have done their part in upbuilding the region and one, Lassen Peak in California, is still active. The mountains have innumerable small lakes of great scenic beauty, of which Lake Chelan in Washington is one of the largest; but there are no vast lakes save Great Salt Lake in Utah, a remnant of a Lake Bonneville of glacial times.

Some of the rivers are extremely powerful, especially the Colorado, which rushes from the Rockies through the Grand Canyon and into the Gulf of California. In Montana, the headwaters of the Missouri supplied a highway for early pioneers to the Columbia, a mighty river that twists through Canada, then flows southwest through Washington and turns for a westward dash to the Pacific. South of Portland is the great valley of the Willamette, a tributary of the Columbia. The Sacramento and the San Joaquin water California's fertile Central Valley. All these rivers have been tamed and turned to the service of man. The Hoover, Parker and Davis dams on the Colorado, Fort Peck on the Missouri in northeastern Montana, Grand Coulee and Bonneville on the Columbia and Shasta on the Sacramento are all great projects completed since 1936. They regulate the flow of the rivers, piping some water off for irrigation, converting some into electric energy; and in the cases of Shasta and Bonneville they have also helped turn the streams into navigable channels.

The Pacific states are broken into narrow valleys that lie between the ranges paralleling the ocean. On the west, in Oregon and California, are the Coast Ranges; on the eastern side of California and extending into central Washington, a generally north-south ridge of higher mountains composed of the Cascade range in Washington and Oregon and the

© EWING GALLOWAY

THE HOPI, sometimes called Moqui, are among the most conservative of all Indians. On their lands in Arizona they still cling to many of their ancient customs. They live in pueblos or villages located on the tops of mesas and cultivate farms in the valley washes below. Their civilization was well-developed long before the white man came. Prehistoric dumps of coal mines have been discovered in Arizona. Apparently the Hopi people at that early period discovered the use of coal as a fuel and found ways to mine it.

WALAPAI WOMEN usually wear simple dresses or skirts and blouses. This one has wrapped a brightly colored blanket around her. These blankets have become quite rare because it takes so long to make them on the simple Indian looms. Most Indians prefer to buy their blankets and clothing from stores in towns close to their reservations. The Walapais live in Arizona, and are included in the Yuman family, which consists of several small tribes now living in the southwestern United States of America and the nearby portions of Mexico.

SHIP ROCK PEAK, a mass of sheer pinnacles, looms against the sky in the Navaho Indian country of New Mexico. It is 7,178 feet above sea level and 1,400 feet above the plains.

Sierra Nevada in California. The latter has occasionally been likened to a long granite block with spurs running westward like the teeth of a giant comb; there is an abrupt drop on the eastern face. Now come three vast high plateaus; that to the north is called the Columbia Plateau, the next, the Great Basin, and that to the south, the Colorado Plateau. As the Cascades and the Sierras rob the winds from the Pacific of their moisture, these plateaus require irrigation where agriculture is practiced. Now come the scattered ranges of the Rocky Mountain system, rising from the Great Plains and the deserts, a part of the rocky backbone of the continent. In it the Continental Divide separates the rivers that flow westward from those that flow eastward.

The Pacific coast is uncommonly regular as compared with the Atlantic seaboard, except for the two huge indentations of Puget Sound—a trough once occupied by glaciers—and San Francisco Bay and the smaller one at San Diego. So narrow is the continental land shelf that there are comparatively few islands save in these bays and sounds, and the United States Coast Guard is obliged to maintain far fewer life-saving stations than on the Atlantic seaboard. A warm ocean current that sweeps northward from Japan, then westward and southward keeps the Pacific coast winters mild; though a narrow belt of cold nearshore water causes fog on the coast, and west of the Sierra-Cascade mountain wall the rains occur in winter. In the higher altitudes of these mountains, however, almost daily thunder showers are precipitated. While the seaward slopes of the Cascades receive exceptional rainfall, from 60 to 120 inches a year, much of the West is semi-arid, and in the extreme south, arid. People who live around Puget Sound wear rain togs all winter as a matter of course, for when it is not pouring, it is likely to be drizzling; but the tourist in southern California is always laughed at for carrying an umbrella on any day that begins merely dull with fog. Parts of Arizona are usually entirely

WHITE-FACED HEREFORD CATTLE pause to drink from a cold stream in Wyoming. The backdrop of snow-clad ranges gives some idea of the splendor and spaciousness of the West.

POTTERY-MAKING is one of the oldest arts among the Pueblo Indians, so called because they live in pueblos, the Spanish word for villages. They make every form of utensil for domestic use, cooking vessels, bowls, platters and candle-sticks. The women are usually the potters, though occasional Hopi men may be found who are also clever craftsmen.

THE APACHES were among the fiercest of all the Indian tribes and were not finally subdued until 1886. This fine specimen has dressed up for the tourist in any finery he could get, no matter how incorrect. The bead ornaments are characteristic of the northern tribes, the headdress is also incorrect, and, of course, no Indian originally had velvet garments.

GENERAL ELECTRIC

ATOMIC ENERGY PLANT AT HANFORD, WASHINGTON

Now a vital spot in world affairs, this area on the Columbia River was desert until it was chosen in 1942 as the location of the world's first atomic energy plant. Hanford had fewer than five hundred inhabitants, but it mushroomed during World War II, until its population reached sixty thousand. Today its buildings cover thousands of acres.

rainless. As for temperature, that varies from the Montana blizzard far below zero along the Canadian border and the eternal snows of the high peaks to the 120 degrees above zero in a city of southeastern California where the writer's heels dented the melting asphalt. The vegetation varies likewise, from the gigantic cedars, spruces and Douglas firs of the moist Pacific northwest, the fog-laved coast redwoods of California and the Big Trees of the Sierra to the prickly plants of the desert and their sage-brush and greasewood.

The Southwest has many palms and other sub-tropic plants, while Washington gardens often look not unlike those of Maine. As for wild life, it still abounds in the remoter regions, although the buffalo that once roamed the plains can now be found only within the boundaries of one or two of the National Parks, and wolves are infrequent save for the little yellow coyotes. A few grizzly bears in the Rockies and a few mountain lions (cougars) in various mountain fastnesses are the only formidable creatures left. There are rattlesnakes in the sunny arid places. Elk and deer are still abundant where there are woods; a few mountain sheep and goats are seen by huntsmen in the northern Rockies and wildcats, rabbits, chipmunks and other small animals may be found in abundance. Western birds migrate up and down the rim of the Pacific and the inland waters are usually alive with trout, besides which salmon by the millions swim up the Columbia at spawning time, actually leaping the falls.

The Spanish explorers early wandered into the interior; and while they made no settlements, the later Spanish influence survives to-day—in the southern half of the region, at least—in many words common in western vocabularies, in place-names and in the modified Spanish architecture, with its patios and loggias of native adobe or tinted stucco or Portland cement. The Indian aborigines varied from the Apaches of desert and plain, who once fiercely combatted the white intruder, to the peace-loving Klatsops, at the mouth of the Columbia, who, in an earlier era dwelt in wooden long-houses with totem-poles before their doors and fished in carved high-prowed canoes, and the peace-loving Pueblos of Arizona who for centuries had practiced a crude system of irrigation.

What we know as New Mexico was visited by the Spanish conquistadores over three centuries ago, and the region has had a continuous Spanish civilization, with few changes, since that date; for not until after the war between the United States and Mexico was American influence felt in that region. The Mission Fathers followed in the wake of the adventurers, and old Spanish Missions are still standing here and there throughout the Southwest. At Mission San Carlos near Monterey lies California's first great missionary, Father Junipero Serra, a Franciscan monk who traveled half the length of the state to bring Christianity to the Indians.

In 1792 Captain Robert Gray, of Boston, sailed in his good ship Columbia into a great river which flowed through what was known as the Oregon Country. This region President Jefferson sent Meriwether Lewis and William Clark to explore. Led by traders of the American Fur Company, which had penetrated as far westward as the Rockies, and by Indian guides, they started in 1804, making their way up the Missouri and across the mountains into the Columbia and keeping such a careful record that it served to guide those who came after them. Now the British George Vancouver, who had served under Captain James Cook, commanded an expedition in 1792 which ex-

THE WAY LOGS ARE HAULED IN OREGON AND WASHINGTON

Timber is the dominant factor in the industrial and commercial life of Oregon. Forest land covers nearly thirty million acres of the state's surface. Eastern Washington and northern California are also heavy timber producers. The "Big Trees" (Sequoia gigantea) of California are shown in the article on National Parks, Monuments and Forests, though not the "Coast Redwoods."

© UNDERWOOD & UNDERWOOD

THIS PUEBLO INDIAN is examining a cane which is said to have been presented to a former chief by President Lincoln. His dress is a strange conglomeration. He is wearing a white man's coat under the blanket which is not one of the tribal designs, nor are the Pueblos supposed to wear such a feather headdress which properly belongs to the tribes of the Plains.

THE RAIN DANCE of the Zuñis in New Mexico is only a part of an elaborate ceremonial which takes place every year around June 21. In recent years some of the pueblo tribes have commercialized their dances by charging admission and inviting the public. Once a year the Zuñis meet with other tribes of the Southwest in an Inter-tribal Indian Ceremony, which is also open to the public. Besides entertaining their guests with colorful ceremonial dances, the Indians bring handmade rugs, pottery and jewelry to the festival to sell to the visitors.

THE MILWAUKEE ROAD

FROM THREE FORKS in southwestern Montana, where the Jefferson, Madison and Gallatin rivers unite, the mighty Missouri begins its 2,547-mile course to join the Mississippi.

plored Puget Sound on both sides of the island of Vancouver (discovered in 1778 by Cook himself). Both the Northwest Fur Company and the Hudson's Bay Company were active in that region.

Thus the Oregon Country came to be claimed by Great Britain, Spain, Russia and the United States. Spain had laid claim to territory on the west coast, but finally by her treaty selling Florida to the United States, agreed to relinquish claim to anything north of what has become the northern boundary of California. Russia, then occupying Alaska, agreed to remain north of 54° 40′, the present southern boundary of that territory. Great Britain desired the Columbia River as the dividing line, while the United States claimed what has become British Columbia, clear to 54° 40′. The two English-speaking nations compromised on the 49th parallel, which gave Washington to the United States. The Oregon Country was gained by right of exploration and settlement, while the Mexican territory was acquired in the war of 1846–47 fought as the result of conflicting territorial claims. A small part was added by purchase.

As early as 1842 John C. Fremont and his guides began a series of explorations of the Rockies, discovering South Pass, one of the three points at which the Rockies can best be crossed (the oldest is that of the Santa Fé Trail). In 1843 a party of settlers, banded together for protection against hostile Indians, made

FILM EDITORS of a motion-picture studio in Los Angeles check prints of a day's "takes." The movie industry has played a leading role in the rapid growth of Southern California.

their way to Oregon in ox-drawn covered wagons, a journey of five months via Fort Laramie and South Pass; and in 1847 the Mormons followed Brigham Young through South Pass to the Great Salt Lake, which had been discovered by Captain Bonneville in 1832. These Latter-day Saints, who even now constitute three-quarters of the church membership of Utah, brought water from the mountains to irrigate their crops and practically made the state. Later those who branched off the Oregon Trail on the way to California left at Fort Hall for what became known as the Salt Lake Trail. To name but one more of the leading steps in the westward course of empire, Captain J. A. Sutter had built a fort in 1839 on a Mexican land grant on the Sacramento River and at his mill, in 1848, gold was discovered in such large nuggets in the gravel of the river bed as led the great migration known as the gold rush of the Forty-Niners. People in all walks of life in other parts of the United States and elsewhere went to California, whether by ship around the Horn, over the Isthmus of Panama or across the plains. It was a race in which the winners were the hardiest or the most acute.

A Spanish post and mission had been established on a hilly peninsula on a land-locked inlet of the Pacific, and in 1835 a town was laid out named Yerba Buena for a small flower that abounded. In 1846 a U. S. man-of-war took possession, the name was changed to San Francisco, and three years later it was a gateway to the gold-mines which drew nearly one hundred thousand people to the state in one year. (The Mission Dolores still stands.)

The harbor filled with the sails of all nations, canvas hotels sprang up like mushrooms, a path over the tide-flats was hastily laid of bales of surplus tea, sand-

SAN LUIS REY is one of the twenty-one missions established by the Franciscan Fathers, led by Junipero Serra (Miguel Jose Serra) between 1769 and 1823. Under their direction, the Indians they had come to convert to Christianity constructed the mission buildings, made furniture, cultivated crops and tanned the hides of their cattle.

A GIANT CACTUS, growing beside the Apache Trail in Arizona, dwarfs the man on horseback. This plant has assumed the form which exposes the least surface to the sun. The pulp is eaten by the Indians, and when squeezed yields a drink, welcome in arid wastes. Desert flowers and blossoming shrubs are also seen along the trail.

lots were auctioned off, the Spanish dons from the ranches were quickly outnumbered by miners in red flannel shirts and knee boots and all was high enthusiasm.

The Santa Fé Trail, primarily a trade route, led from Independence, Missouri, eventually to Los Angeles, and the ponderous freight-wagons had worn a way both wide and deep in the sun-baked sands. There followed the horse-stages and for two romantic years preceding the first telegraph line, the pony express. In time bands of steel rails were flung across the continent, and while they were extremely costly, they laid the way to more rapid settlement. Where herds of buffalo had sometimes stopped the trains, the cattle country in time gave way to fields of waving wheat, and later to oil derricks or mine mouths with debris like giant anthills; and today the luxurious transcontinental trains offer one radio music while the scenery whizzes by. The Moffat Tube under James Peak penetrates the Continental Divide fifty miles west of Denver, rising to an elevation of nine thousand feet; and there is an electrified tunnel through the Cascade Range a hundred

GIANT HOOVER DAM KEEPS THE COLORADO RIVER UNDER CONTROL
The tremendous power plants below Hoover Dam look like toys when seen from the top of the dam. But if you go down to the plants through one of the towers, you will find the roaring machines that provide electric power for much of California, Arizona and Nevada. Blocked by Hoover Dam, the Colorado River forms Lake Mead, the world's largest artificial lake.

CHARLES MAY

PATIO AND SWIMMING POOL of a luxury hotel on the "strip," in Las Vegas, Nevada. The "strip" is a road leading out of the city, lined with beautiful homes and palatial hotels.

miles east of Seattle. Now great national highways parallel the railroads across deserts and mountains and from Vancouver down the coast to San Diego. The considerable distance of San Francisco and Los Angeles to other population centers made California an important state in the dramatic history of the airplane.

And what of the people who make up the West? While in 1950 the three Pacific states averaged 45 persons to a square mile, the eight Mountain states averaged just about six, that is, 5.9, including Indians; Wyoming has but 2.9 persons to a mile and Nevada averages 1, as compared with 300 of the Middle Atlantic states and the average of 50.6 for the entire United States. Yet the area of the Mountain states is 863,887 or between a third and fourth of the continental United States. Including the Pacific states, the West totals 1,187,753 square miles.

Even today there are mountain and desert regions where one may ride for days without seeing a human habitation, although in other places one finds thronged motor highways, vertical architecture in the downtown districts and shop windows which display the latest styles from the fashion centers.

The stricter laws put through in the 1920's cut off immigration from abroad to a thin trickle. Earlier, however, many newcomers came from northwestern Europe, Canada and Italy and there were a few from China and Japan. People from the Scandinavian countries were particularly attracted to the big lumbering operations of the Northwest. Italians were drawn to the vineyard slopes—reminding them of their sunny homeland—

of California. Seasonal workers flooded across the border from Mexico—some of them illegally. This situation became a serious problem after World War II. For every worker who came in officially there were perhaps five "wetbacks"—as the workers who enter illegally are called. Some of them swim or wade the Rio Grande, hence the name. In an effort to control matters, Mexico and the United States agreed in 1955 to provide more effective border patrols and better living conditions for legal entries.

We can see that in a region of such varied geography, there must be a number of ways for people to earn their livings. Agriculture is possible in certain areas, only by irrigation; but stock-raising may be practiced over large expanses, and where range-cattle cannot find a living, there may be enough forage for sheep, which are able to go for longer periods without water. In the forested regions, lumbering is the leading industry; along the coast and the Columbia, fishing is important, and in the region around Los Angeles, the sinking of oil wells and refining of petroleum take high rank, for California is one of the most important oil states of the Union. The milling of the lumber, the canning of salmon, fruit and vegetables and the milling of wheat flour are to be expected, and California refines Hawaiian sugar. The sunshine of southern California also makes possible a gigantic moving-picture industry, and airplane factories cluster around Los Angeles. In the slopes of the Rockies, mining for copper, silver and lead is of tremendous importance.

Where People Own Their Farms

In this division of the country, farms are, in the majority of cases, worked by owners though tenancy is increasing rapidly. The proportion of tenancy in some of the Mountain states is somewhat less than in the Pacific states. The farms of the Mountain states are on the average somewhat larger than those of the Pacific states, though there are many great tracts under one management in California particularly. Many years ago tall stories were told of the "bonanza farms" so large that a man could plow only a furrow or two in a day. There are now some enormous farms worked almost entirely by machinery where tractors pull many ploughs and wheat is reaped and threshed in one operation by combines. On the other hand, there are many small farms and orchards upon which the farmer's family does all, or nearly all, of the regular work.

On many of the farms of these Western states, wheat and hay are the principal staple crops, aside from the fruits and vegetables of certain limited regions. Washington recently has run second to Montana in its wheat crops with Idaho and Oregon next. These two groups of states usually produce something less than a fourth of the total wheat crop. The proportion of hay produced in the section is about the same as of wheat, that is between a fourth and a fifth.

A Famous Fruit Region

One is likely to think first of California fruit, but some years California's leading single crop is hay. In that mild, dry climate successive crops of alfalfa may be raised and the hay of the state all put together amounts to very nearly five million tons a year. But California produces millions of boxes of oranges, to say nothing of apples, peaches, pears, lemons, prunes, apricots, figs and grapes—wine, table and raisin. California truck-farms produce millions of dollars worth of lettuce alone. The Inland Empire east of the high ranges in Washington and the Hood River Valley of Oregon are two regions which raise quantities of fine apples. Oregon is also famous for its berries, especially loganberries, which spring up quickly when planted on cutover or burned lands and have been known to send out runners fifty feet in length the first season. California's long interior valley between the Sierras and the Coast Range is one of the richest in the world and the Imperial Valley in the extreme south is fertile. But in much of the West irrigation is necessary and Colorado, for one, has an extensive system of canals which have long been the property of the

A COG RAILROAD takes sight-seers up Mount Manitou Incline, which is near Pikes Peak. From the top of the mountain, there is a sweeping view of the eastern ranges of Colorado's Rockies.

state. Since 1902 a Federal Reclamation Service has established irrigation projects where, by intensive cultivation, even deserts have been made to yield food for men or cattle, although sometimes the costs exceed the productive capacity of the land.

Cattle-ranching Important

On the vast ranges of these states are millions and millions of cattle. Colorado has the largest number among the Mountain states with Montana next while California leads among the Pacific states. There are some great ranches though most such establishments as we read about in stories of the Old West have been broken up. Much of the public lands on which the cattle ranged has passed into private hands, and cattle are no longer allowed to roam at will in the national forests.

At one time the wild horses appeared to be on the point of disappearing, but their numbers seem to have increased. In some regions they have become a pest, as they eat the grass needed for cattle and themselves have little value. There are half-wild horses upon ranches which are broken with difficulty. This was the origin of the rodeos which are now held as a spectacle to which tourists and others pay admission. Some of the horses are trained to buck and rear to excite the wonder of the tenderfoot. Montana and Colorado have the greatest numbers of ranch horses but motor cars, motor trucks and tractors are being substituted for horse power both in town and country. Strange to say, many people still believe that the wild horses are native to America. As a matter of fact, they are descendants of the Spanish horses which escaped from the early explorers, and also from many strays which have run away from their owners and joined their free kindred.

Where the forage is too scanty for cattle, sheep can find a living. Almost half of the sheep in the United States are in these two groups of states. Montana is usually first with California next but all these states have many sheep and are extensive producers of wool. Wyoming leads in wool production with Montana second and California third.

Wealth of Mines and Timber

There is more lumber produced in the Pacific states than anywhere else in the continental United States—in fact, 44.9 per cent of the whole amount (besides wood pulp, shingles, turpentine and rosin). The Pacific states are producing more than fourteen billion board feet per year. Unfortunately, the dry summers see many destructive forest fires.

While the fisheries produce much wealth, the mines of the Western states are the largest producers of wealth. The figures of mineral production are stupendous, though gold and silver no longer hold their proud positions. Some of the commoner metals now have a greater annual value than that of gold and silver. Copper, for example, has a much greater value than gold. Arizona is the great copper state, followed by Utah and Montana though all of these states produce more or less of this useful metal. In fact, Michigan is the only important producer of copper outside of these two groups of states. The humble lead is worth more than the gold. Idaho leads the West, with Utah, Arizona and Montana following.

Petroleum and natural gas are the most valuable mineral products of the Western states as of the country as a whole. California is the chief producer though Wyoming and New Mexico furnish appreciable quantities and several other states produce less. California, in fact, is one of the chief oil producing states in the nation.

Mineral Production in the West

The Western states produce about three-fourths of the gold and nearly all the silver mined in the United States proper. California is first in gold with Utah next, in the section (South Dakota in another group is actually first). Idaho is easily first in silver with Montana generally second though in some years Utah is ahead. Arizona, Colorado and Nevada are also large producers of silver. Much of the silver secured is a by-product from mines worked for copper, lead or zinc.

THE STATES TOWARD THE SUNSET

Western educational facilities are good. Every state has a state university. Men from all of the Mountain states attend the University of California, with its main headquarters at Berkeley and a branch at Los Angeles. The university is by far the largest in the West and in point of numbers ranks as one of the three largest in the United States. Its southern branch includes Scripps Institution of Oceanography at La Jolla and Lick Astronomical Department at Mount Hamilton, where the dry, clear air is wonderful for observing the stars. Other well-known astronomical observatories in California are at Mount Wilson, near Pasadena, and at Palomar, near San Diego. Palomar's 200-inch telescope is the world's largest. Stanford University, near Palo Alto, has been provided with a remarkably large endowment, over $39,000,000, much of it given in memory of the boy whose name it bears. The University of Southern California has more than 17,000 students enrolled. There are many denominational and independent colleges. The Mormon Church in Utah maintains the Brigham Young University and the Latter Day Saints University, which has a business college and night school. But there is not space for mention of all the educational facilities of the West. There is a general desire for education, and every state has compulsory education. There is a good showing of every kind of school from primary to normal, state university and private institution, especially in proportion to the population.

Hardy, self-reliant, brave to the point of daring were those first pioneers into the unknown West. The virtues of the times—which the survivors handed in large degree to their progeny—were courage and enterprise, generosity to those in need and summary justice for the fugitives from eastern justice and other "bad men" who showed no respect for life and the means to life, notably, horses. Toward the few women of the early days there was a gallantry which made them safe to a degree seldom met in the world's history. Initiative and creative energy still stamp the western mind, as does an intellectual independence and readiness to

Courtesy Reno Chamber of Commerce

PYRAMID LAKE SPIRES, SET IN THE RUGGED TABLELAND OF NEVADA

Most of Nevada, a huge state of but little more than 160,000 population (including Indians), lies in the Great Basin from four to five thousand feet above the level of the sea. Broken by buttes and high ranges, its valleys contain a few lakes like those above, and more mud lakes up to fifty miles in length but only a few inches deep, which evaporate in summer.

experiment on a large scale. As one outcome, certain of the western states were the first to grant equal suffrage. A fondness for exploration and a readiness to rough it has led to an unusual degree of camping and mountaineering. Local patriotism runs high. There are historical pageants like that at Eugene, Oregon, a few years ago, which displayed covered wagons and the Indian "dug-out" canoes. There are flower festivals—tulips in Washington, roses in Oregon, and on every New Year's Day in Pasadena a so-called Rose Tournament in which thousands of flowers of every kind are wound about floats from nearly every town in southern California, before something very like a Spanish fiesta. California celebrates not only Admission Day, the anniversary of the date when the state was admitted to the Union, but every New Year's Eve holds a Mardi Gras. There is cheerful rivalry among the smaller cities, and the standard joke of cosmopolitan San Francisco is to ask anyone from Los Angeles, "What part of Iowa did *you* come from?"

Parts of the West still contain isolated communities reached only by horse-drawn mountain stages, which up to the turn of the century were common almost everywhere in the country districts. Travelers wore linen "dusters"—for good reason— and so few were the places on the winding mountain roads where two vehicles might pass that the driver watched for the approaching spiral of dust which betokened an oncoming team. In remote valleys people exchanged their surplus produce and lent mutual aid at house-raisings, and entire families drove long distances to parties, where the grandmothers put the babies all to bed in some one room. It was a friendly West—and still is, outside the larger cities. Even the cities are not quite like those of the East.

THE WEST: FACTS AND FIGURES

STATE	TOTAL AREA (SQ. MILES)	POPULATION (1950 CENSUS)
Montana	147,138	591,024
Wyoming	97,914	290,529
Colorado	104,247	1,325,089
New Mexico	121,666	681,187
Arizona	113,909	749,587
Utah	84,916	688,862
Idaho	83,557	588,637
Nevada	110,540	160,083
California	158,693	10,586,223
Oregon	96,981	1,521,341
Washington	68,192	2,378,963

COMMERCE AND INDUSTRIES

Agriculture and mining are the more important occupations throughout the western states. With the development of irrigation, the Pacific coast states have become the leading source of supply of the nation's fruits and vegetables. California, Oregon and Washington produce enormous quantities of apples, peaches, prunes, pears, citrus fruits, apricots, cherries, berries of all kinds, nuts and vegetables which are either dried, preserved or shipped fresh. Other important crops are wheat (Montana and Washington); hay (California, Montana and Oregon); sugar-beets (Colorado and Utah); cotton (California and Arizona). Cattle and sheep-ranching important in all of the western states; most of the country's wool supply comes from this region. Over three-fourths of the national forests are located in these states; Washington and Oregon lead in the amount of timber cut each year. Extensive fisheries occur along the Pacific coast. Mineral products are rich and varied; over 98 per cent of the silver produced in the United States comes from the western states (Utah, Montana, Idaho, Arizona and Colorado lead); about 75 per cent of the gold (chiefly in California, Utah and Nevada); and over 95 per cent of the copper (Arizona, Utah, Montana and New Mexico lead). California is a leading source of petroleum; asbestos, lead, zinc and wolfram (tungsten) are mined in several of the states. Manufacturing increases, especially in coastal metropolitan areas; leading manufactures are lumber, aircraft, nonelectrical machinery, canned, preserved and frozen foods, motion pictures, chemicals and petroleum products.

IMPORTANT CITIES

Population of state capitals (1950 census): Phoenix, Arizona, 106,818; Sacramento, California, 137,572; Denver, Colorado, 415,786; Boise, Idaho, 34,393; Helena, Montana, 17,581; Carson City, Nevada, 3,082; Santa Fe, New Mexico, 27,998; Salem, Oregon, 43,140; Salt Lake City, Utah, 182,121; Olympia, Washington, 15,819; Cheyenne, Wyoming, 31,935.

Population of other important cities (1950): Los Angeles, California, 1,970,358; San Francisco, California, 775,357; Seattle, Washington, 467,591; Oakland, California, 384,575; Portland, Oregon, 373,628; San Diego, California, 334,387; Long Beach, California, 250,767; Spokane, Washington, 161,721; Tacoma, Washington, 143,673; Berkeley, California, 113,805; Pasadena, California, 104,577.

SUN AND SNOW and the Swiss-chalet look of the buildings make a winter wonderland of Sun Valley, Idaho. Skiing is a popular sport of many mountain sections of the West.

VAST FIELDS OF CROPS, as well as grazing lands, spread across the broad valleys and Plains of Montana. Hay, fodder for livestock, is cut and stacked with modern machinery.

PHOTOS, STANDARD OIL CO. (N. J.)

RODEO DAY is the big event of the year in many a mountain-state town. Horse races, trick riding, broncobusting and steer roping are among the most popular cowboy contests.

ASAHEL CURTIS

AWE-INSPIRING MOUNT RAINIER VIEWS ITSELF IN MIRROR LAKE

The ice-clad volcanic peak rises more than fourteen thousand feet above sea level in the Cascade Range in the state of Washington. The Indians named it Tacoma, "the mountain that was God." Mount Rainier National Park, with its deep conifer forests and glaciers, attracts motorists and campers; and the towering peak challenges daring mountain climbers.

MOUNT OF THE HOLY CROSS, in Eagle County, Colorado, was named from the cruciform lines of snow in the ravines near its peak. The mountain is 13,996 feet high, the upright shaft of the cross measuring two thousand feet and the transverse ridge forming the arms, eight hundred feet. The State of Colorado, lying at the junction of the Great Plains and the Rockies, has many mountains over 12,000 feet high, while about forty peaks reach to a height of over 14,000 feet. There are lofty plateaus between the ranges, as elsewhere in the Rockies.

MAJESTIC CLIFFS like these line the Colorado River along its route through the mesa country. The 1,700-mile river starts in the mountains of north-central Colorado and winds southwest through wilderness, plateaus and deserts, flowing at last out into the Gulf of California. In cutting its way through the cliffs and canyons, the Colorado has literally carved out geological monuments to the past. The layers of rock exposed in these cliffs are really a map of the crustal warpings, wind erosions and volcanic eruptions that took place millions of years ago.

THE AMAZING EFFECT of sweeping curves and mountain terraces results from open-pit copper mining at Bingham, Utah. Trains chug along the various levels, carrying out the ore.

INDIANS of the Northwest, like generations of their forefathers, net and spear the big salmon that fight their way over rapids and falls of the frothy Columbia River to spawn.

© GIFFORD, COURTESY UNION PACIFIC SYSTEM

MULTNOMAH FALLS, dropping over the edge of a cliff 850 feet high, form part of the beauty of the scenery along the Columbia River Highway, which skirts the southern shore of the river, following the course of the old Oregon trail through the forested Cascade Mountains. It also passes Latourelle, Horsetail, Mist and Bridal Veil falls.

UNION PACIFIC SYSTEM

CALIFORNIA GARDENS count among their treasures shrubs, foliage plants and blossoms that represent the flora of many climes; and here nature intensifies their growth until rose-bushes become rose-trees and bloom lavishly up to Christmas, lemon-verbena aspires to tree proportions, heliotrope develops into hedges breast-high, and colorful nasturtiums sometimes cover an embankment in the place of grass. Cannas unfold great leaves and flowers, water-lilies grace the pools, and pepper-tree branches droop with clusters of decorative crimson berries.

WEIRD TOTEM POLES on the main street of Juneau, Alaska, are reminders of the native Indian culture. The city is in the heart of one of the world's great gold-producing regions.

Lands of Treasure and Romance

United States Territories of Hawaii and Alaska

In this article we tell you about two of the most important lands under the American flag. Physically, they have only two things in common. Both are to the west of the United States proper and both are washed by Pacific waters. In most other respects, no two regions could be more unlike. One is a chain of islands, in the path of refreshing trade winds, where it is summer all year round. The other is a part of the continent of North America, extending far within the Arctic Circle. Each is beautiful, but one has an enchanting loveliness, and the other a majestic grandeur.

MANY of the great states of the Union once were known as "territories." This meant that they belonged to the United States and yet could not participate fully in self-government. Their people had no vote for president, and their governors were appointed by the White House instead of being elected at the ballot box. Furthermore, a territorial delegate in Congress could deliver speeches but was not allowed to participate in roll calls on bills. He had no real power and generally not much influence.

Such famous areas as Missouri, Oregon and Washington occupied territorial niches before being admitted as sovereign states. The old Northwest Territory was subdivided into the important states of Ohio, Indiana, Illinois, Michigan and Wisconsin.

Today only two such territories remain. They are as unlike as any two realms could possibly be. One is Hawaii, a chain of eight semitropical islands 2,000 miles west of San Francisco in the Pacific Ocean. The other is Alaska, a vast domain 1,000 miles north of Seattle, sprawling across the cold gables of the planet.

In the mind of the average person Hawaii is typified by pineapple, sugar-cane and girls swaying to the strum of ukuleles. Alaska calls up vivid images of Eskimo families in fur parkas, of bushy sled dogs, of salmon canneries and of great brown bears weighing more than half a ton.

Yet these two contrasting territories are united by a common bond—a burning, zealous desire for statehood. Referendums among the voters in both Hawaii and Alaska have decisively favored admittance to the Union as the forty-ninth and fiftieth states. Hawaii has been a territory since 1900. Alaska was purchased from Russia in 1867, and a territorial government came into existence in 1881.

One of the strongest desires of each resident of these two colorful territories is to count himself entitled to a ballot in presidential elections and to genuine representation in Congress. These aspirations will automatically be realized when Hawaii and Alaska become states.

Hawaii

The Hawaiian Islands are one of the few places under the American flag where the same clothes can be worn Christmas Day as on the Fourth of July. In Honolulu, capital city of this picturesque territory, the average temperature for the month of December is 73 degrees. During midsummer, in July, it is barely 5 degrees warmer—77.9.

These steadfast temperatures, always balmy, give to life in "the Islands" a languid, restful quality. Bathing in the tepid ocean is the most frequent recreation. Newcomers are surprised to find hard coral beaches rather than soft sand underfoot. At anytime of the year a hostess can pile her guests' plates high with luscious fruit from her own garden—pineapples, avocados, bananas.

People in the Hawaiian Islands rarely hurry. The big ships from the mainland move slowly into Honolulu harbor, and native boys dive nimbly for pennies thrown overboard by the passengers. Brown-

COURTESY MATSON NAVIGATION COMPANY

A NATURAL BRIDGE adds interest to this inlet on the shore of the island of Hawaii, a part of the coast that has abrupt cliffs of volcanic formation, instead of gently sloping coral beaches. On the same island, which has given its name to the whole group, a driveway between ferns of tree height leads to the part of Hawaii National Park where the periodically active volcano Kilauea is set "on the hip" of the older volcano Mauno Loa. From the brink of the crater one can look into the pit of fire where, legend says, dwells Pele, goddess of the volcanoes.

COURTESY MATSON NAVIGATION COMPANY

WAIKIKI BEACH, on the island of Oahu, with its bordering coral reefs, is an ideal spot for water sports. The temperature of the water makes swimming and surf-board riding delightful. At night, when the ripples and the palm trees are silvered with moonlight, a tropic enchantment holds sway. Beyond the curve in the shore we can see the crater Diamond Head.

HAWAIIAN PINEAPPLE CO.

PRICKLY GOLD

This luscious fruit represents a $70,000,000 industry that employs thousands of workers.

working Hawaiians is on a sugar payroll Pineapple, a symbol of Hawaii to all the world, ranks next among the industries of these verdant islands. The first seafaring men on the broad Pacific reveled in the taste of Hawaiian pineapple. In 1951 Hawaii shipped to the mainland 24,000,000 cases of pineapple and pineapple juice. This sweet fruit with its distinctive flavor was worth $88,000,000 in that year. Most

HAWAII VISITORS BUREAU

TALL AND SWEET

Hawaiian plantation workers display their skill in harvesting a field of juicy sugar-cane.

skinned girls weave gently in their famous hula dance, which is done more with the hands than with the hips and which has a symbolic meaning deep in Hawaiian tradition. The "grass skirt" worn by the dancers is not made of grass at all but from fresh ti leaves. The ti plant sprouts a long, dark green leaf that resembles the leaf of the banana tree.

In spite of the languor of Hawaiian existence, the 499,794 people of the islands have created a prosperous economy. Indeed, the Federal Government at Washington, D. C., collects more money in taxes annually from the residents of the Hawaiian Islands than from the taxpayers of any of eleven of the states of the Union with smaller populations. In addition, advocates of statehood emphasize the fact that Hawaii now has more inhabitants than any state when it was admitted to the nation except Oklahoma. Hawaiians are civic-minded. Eighty per cent of Hawaii's registered voters participate in territorial legislative elections.

Sugar-cane is the basic industry. Production of sugar in 1951 totaled 960,000 tons and this immense crop was valued at $124,000,000. One out of every eight

of the companies now processing this zestful fruit have existed in one form or another since Americans first came to Hawaii.

Sugar plantations and waving fields of pineapple are expected in Hawaii. The islands' cattle ranches are not so generally known. Yet Hawaii produces $7,000,000 worth of beef products a year. White-faced cattle graze on lush uplands which look down dizzily on the blue sea. These ranches are patrolled by hard-riding cowboys, as dextrous and accomplished on horseback as any in Arizona or Texas.

SURF, SUN AND CORAL SAND AT WAIKIKI BEACH

Tropical vacation paradise of the Pacific is Hawaii with its fabulous beaches and clear waters so popular with surf-board enthusiasts. Diamond Head on Oahu Island rises in the background.

An extraordinarily large proportion of the fertile land of the islands is devoted to agriculture. The farms of Hawaii yield $2,700,000 worth of coffee beans annually, in addition to diversified dairy products for the ever increasing population.

The term "Hawaii" is all-inclusive. It refers to the eight islands that form the Hawaiian group; yet the name actually comes from the single island of Hawaii, largest in size of the chain. The Hawaiian Islands together have a total area of 6,454 square miles. This is comparatively small as states of the Union run, yet it exceeds the respective areas of Rhode Island, Delaware and Connecticut.

Strangely enough, it is not the dominant island of Hawaii that is the best-known in the Hawaiian group. This honor is reserved for the third largest island in area—Oahu. Honolulu, with half the people of all the islands inside its municipal boundaries, lies on Oahu. And Oahu is the site of the great American naval base of Pearl Harbor. It was here, on December 7, 1941, that occurred the act which the late President Franklin D. Roosevelt said would "live in infamy." When Pearl Harbor was bombed without warning by Japanese planes, the United States was brought into World War II as an active belligerent.

Many Japanese-Americans live in the Hawaiian Islands. They formed the herioc 442nd Battalion, which fought through the Italian campaign and made a brilliant record in battles and campaigns. In spite of the presence of numerous residents of brown skin and Oriental extraction, Hawaii never has been seriously plagued with racial troubles. Democracy has attained a high level in the islands and there is little discrimination. In 1950, people of many ancestries took part in a

PHOTOS, HAWAII VISITORS BUREAU

A WARRIOR KING

Before Honolulu's Judiciary Building stands a statue of King Kamehameha I, the Great.

FROM NUUANU PALI, "place of vision," stretches out a stupendous panorama of sea and shore and volcanic headlands. It is part of an old crater and may be reached by a drive from Honolulu along the southeast shore of Oahu. A motor ride all the way round the island on the excellent roadways affords a great variety of scenery not only of mountains and beaches but of cultivated landscape as well, in the sugar and rice and pineapple-growing districts, where the soil has by scientific methods been brought to a very high state of fertility.

COURTESY PACIFIC STEAMSHIP COMPANY

SITKA, ALASKA, seen here from Sitka Sound, has a safe and spacious harbor. Until 1867 the city was the seat of government of Russian America, and here, on October 18, 1867, an historic event took place. The flag of Russia was hauled down from the government building and the Stars and Stripes run up. Sitka remained the center of government until 1906 when Juneau was made the capital of the United States territory. In the old Russian cathedral of St. Michael, there is a valuable collection of art treasures and objects of historic interest well worth visiting.

SCHOOL DAYS IN THE TROPICS

The modern style of the administration building of the University of Hawaii suits the climate.

convention to draft a constitution for the prospective state of Hawaii. The document was modeled after the constitutions of the United States and of the state of California.

Visitors to these islands always are amazed by the great variety in geography. These "lily pads" in the Pacific do not conform to the notion of tropical islands as flat places. The island of Kauai, for example, is split by awesome Waimea Canyon, a deep chasm that rivals any abyss on the mainland except the Grand Canyon of the Colorado. Mauna Loa, a mountain battlement on the island of Hawaii, soars to 13,784 feet. Mauna Loa is actually more massive in circumference than either Mount Whitney in California or Mount Rainier in Washington.

Sun visors, shorts, bathing suits, toeless sandals, beach umbrellas—these are the standard garb and equipment. Few days pass without sunshine. A savage rain squall may drench Honolulu fiercely for half an hour, but then the sun will pierce through the overcast and soak up the water that has fallen to earth. Annual rainfall in the capital city is rarely more than thirty-two inches—less than in New York or Seattle or Portland, Oregon.

In the islands people of Japanese ancestry outnumber the Caucasians. Yet the boys and girls of these families are educated together in a modern public-school system, and the relation between races is one of comradeship. In Hawaii 94,000 children are enrolled in public schools and 24,000 in private schools. Teachers are well paid, and there is always a long waiting list of men and women from the mainland who hope to voyage to Hawaii to teach. The University of Hawaii offers degrees in scientific and liberal arts subjects, and its athletic teams compete frequently with colleges from "the States." Spectators at football games invariably marvel at Hawaiian punters who can kick a ball fifty yards with their bare toes. They forget that these young men have run across the firm coral beaches without footgear almost since they could toddle. The University of Hawaii is a land-grant college like many of the schools in states west of the Mississippi River.

Under territorial status, Hawaii's government is largely a patchwork. The governor is appointed by the president of the United States, but the people of the islands elect a legislature consisting of fifteen senators and thirty representatives. Yet Congress can set aside any law passed by the Hawaiian assembly. When they

PHOTOS, HAWAII VISITORS BUREAU

HAWAIIAN COUTURIERES

No souvenir of Hawaii is more popular than the "grass" skirt these girls are making of ti leaves.

LANDS OF TREASURE AND ROMANCE

contemplate the fact that they pay more than $81,000,000 a year in federal taxes, the residents of Hawaii feel that they are entitled to more self-government than so far has been allowed them.

Hawaii lays a claim on the heart of its visitors. As a great liner puts out to sea from Honolulu harbor, the people on shore sing ALOHA to their friends. The plaintive strains of Hawaiian music come across the smooth water. *Aloha* means "good-by" in the lilting language of the native people of the islands. The hard, stern realities of life must be faced again. The pleasant and placid routine of Hawaii fades on the horizon. Jack London once wrote that anyone who spent a month in these sunny islands was sure to return before he died.

Nenana during the long summer days, bare-chested men frequently load riverboats in sweltering temperatures that touch 100 degrees above zero. Few places on earth record such startling contrasts.

Alaska's forests contain seventy-eight *billion* board-feet of timber. These trees, harvested on a basis of perpetual yield, could supply at least 25 per cent of the newsprint needed on the continent. Early in the 1950's, construction began on the first pulp mill in the North, near Ketchikan, Alaska's principal salmon-packing center. Other mills are eventually planned for Sitka, Juneau, Wrangell and Petersburg. The names of some of these communities indicate the Russian influence in the settlement of Alaska.

ALASKA DEVELOPMENT BOARD

THE BUSH PILOT DOES HIS PART
The busy Alaskan bush pilot carries both mail and passengers long distances. Here a pilot has landed his plane, equipped with pontoons, on the Taku River, in southeastern Alaska.

Alaska

Alaska, land of contrasts, ascends to the 20,300-foot summit of Mount McKinley, loftiest spot in North America, and it has a longer coastline than the United States proper. Alaska includes lonely valleys that have never been mapped and bustling cities where a vacant room is rarely available. Winter temperatures often drop to 75 degrees below zero, with anti-freeze solution hardening in automobile radiators. Yet despite its latitude, only 3 per cent of Alaska is permanently blanketed with ice and snow. At Fort Yukon or

From 1940 to 1950 the population of Alaska shot from 72,000 people to more than 128,000. This is a 77 per cent increase, largest anywhere under American sovereignty. Most of the recent newcomers have migrated northward to Anchorage and Fairbanks to work on defense projects. A few years ago Anchorage was a tiny outpost. Now it has housing developments, two newspapers, bus routes, department stores, wide avenues and many hotels. These burgeoning Alaskan communities are inhabited by two or three men to every woman. A pretty nurse or school teacher arriving from the states is barraged with early proposals of marriage.

However, local authorities advise women not to migrate to Alaska until they are sure of a place to live. Residential quarters often cannot be obtained for any price.

If Alaska achieves statehood, Texas will lose its claim to first in size. Alaska measures 586,400 square miles, approximately twice the size of Texas. In addition, Alaska extends so far from west to east that it lies within three different time zones, which are spoken of in the north as "Yukon time," "Alaska time" and "Aleutian time." At its longest and widest points, if we include the great Aleutian archipelago, Alaska actually is longer and wider than the United States itself. During World War II the Aleutian Islands became the first North American soil (north of the Rio Grande) to feel an invader's tread in nearly a century and a half. Imperial Japanese troops landed on Kiska and Attu but later were expelled by American soldiers fighting in damp cold and chilling fogs.

At the time of the Aleutian invasion, a

TRIPLETS ON A TOTEM POLE

These curiously carved figures decorate the totem pole of a chief, on Wrangell Island.

1,523-mile highway was pushed through the spruce solitudes, as an emergency route to Alaska, by seven regiments of the United States Army Engineer Corps. Later the road was widened and steel bridges were built by civilian workers. Until the completion of this highway, Alaska was to all intents and purposes an island—as cut off by the Canadian wilderness as an island is by the sea. But in 1951, for the first time in history, more people migrated to Alaska by land than by boat. These newcomers arrived over the Alaska Highway. Out of the total of 22,507 men, women and children who drove to Alaska on the gravel road, more than 5,880 stayed on to make their homes.

The airplane also has had a profound influence on modern-day Alaska. Once a wilderness family was cut off completely after the first "big snows" had fallen. A serious illness might mean death because no doctor could get through drifts as high as a lodgepole pine. Now a plane on skis can land almost anywhere that there is open country. The "bush pilot" is to the Alaska of the 1950's what the dog-musher was to the Alaska of 1898, when gold was

OLD RUSSIAN CHURCH AT KODIAK

This church, with its Russian-style cupola, is a reminder that Alaska once belonged to Russia.

found in the Klondike and in other murmuring creeks of the north.

Alaskan cities now get home delivery of mail. This, too, is directly attributable to the airplane. The old mail schedule of one or two boats a week never justified the maintenance of a force of letter carriers. As a result, all Alaskans were compelled to pick up their mail at the post office. Anchorage had more central-station post-office boxes than even New York City. Now the plane comes with mail every day and so the mailman in his familiar gray uniform is justified at last.

Fisheries are still the principal industry. Alaska produces 88 per cent of the canned salmon sold in American stores. The annual salmon pack is worth about $60,000,000. Many Alaskans, both Indian and white, troll for salmon offshore in small boats. The large king salmon, or *tyee,* is caught in the southeast and the smaller red salmon in Bristol Bay, near the Bering Sea. Many salmon are snared in fish traps, huge timbered chambers fitted with nets and placed near the mouths of rivers where salmon spawn. Most Alaskans believe these devices threaten the future of the fisheries and they have petitioned Congress to outlaw the traps.

It costs a great deal to live in the territory, even by present-day inflationary standards. A dozen eggs may sell for $1.50, a quart of milk for 35 cents. Haircuts are $2.00 in Fairbanks. Soldiers at Alaskan bases during the war were accustomed to paying 75 cents for a milkshake. Wages are correspondingly high and a carpenter helping to erect a barracks may earn $750 a month. It is often easy come, easy go. Thousands of dollars are wagered on the exact day, hour, minute and second when the ice pack will drift out of the great Tanana River. Perhaps this attitude toward money is due to the preponderance of men and the absence of a hearth and home life.

Tuberculosis has been a particularly grim problem. Some Indian and Eskimo tribes are wracked by the world's highest death rate from tuberculosis, higher even than in famine areas of India and China. For many years nothing was done about this. However, not long ago a governor of Alaska, Ernest Gruening, himself a medical doctor, established Alaska's first

BLACK STAR

JUNEAU, THE CAPITAL OF ALASKA
Juneau, in southeastern Alaska, on the Gastineau Channel, is the main port of the territory. It nestles between the water and the slopes of Mount Juneau and Mount Robert, rising behind.

full-time Department of Health. Hospitals and clinics are being built. A converted Navy vessel, now called S. S. Hygiene, cruises into fiords taking chest X rays of native families. Infected adults and children are immediately isolated from healthy tribespeople so the plague will not spread farther.

Alaska is a lively, tingling realm. Its people are typical frontiersmen. They face danger bravely and seldom quail from hardships, even in sports. Alaskans compete in dog-team races with scarlet-coated Canadian Mounties from Whitehorse across the border. A high-school basket ball team may fly eight hundred miles to oppose the team in the next town. This is equivalent to the distance between New York and Chicago. Alaskan children trudging through the woods to school have been taught to freeze into silence if they hear an ominous growl. It may come from a Kodiak brown bear, the largest meat-eating creature that stalks the earth. Many youngsters catch a fifty-pound salmon before they weigh much more than that themselves.

Indians and Eskimos comprise 20 per cent of Alaska's total population. At one time these people were the victims of discrimination and could not share equally in the use of hotels, theaters and other public places. But in 1945 the territorial legislature at Juneau enacted a civil-rights bill ending such abuses. Now six natives sit as members of the legislature, and a full-blooded Tlingit Indian has served as president of the Alaskan Senate.

Juneau, the capital, is the farthest north seat of government in the Western Hemisphere. It lies on a narrow lava shelf between the salt waters of Gastineau Channel and the great sheer rock precipices of Mount Robert and Mount Juneau. Old Greek Orthodox churches, relics of the Russian occupation of almost a century ago, contrast vividly with the marble pillars of the Federal Building, where the American flag flies. In summer, when the hours of darkness are short, Juneau's people play tennis, hike up the towering mountain and swim in the icy waters fed by near-by glaciers. If statehood comes, with its exciting elections and self-government, gossip about politics may be added to the pastimes of this capital in the glow of the midnight sun.

By RICHARD L. NEUBERGER

ALASKA AND HAWAII: FACTS AND FIGURES

Alaska includes extreme northwestern part of North America and the Aleutian Islands. Bounded on the north by Arctic Ocean, south and southwest by Gulf of Alaska and Pacific Ocean, west by Bering Sea and Strait and east by Yukon Territory and British Columbia. Total area, 586,400 square miles; population (1950), 128,643. Governed conjointly by the United States Congress and by local legislative assembly of 16 senators and 24 representatives. Governor and judges appointed by the president of the United States. The raising of fur animals, fishing, forestry and mining are the principal industries. Exports are salmon, herring, copper, furs, shellfish and gold; imports are iron and steel, tin cans, machinery, mineral oil, meats, provisions and explosives. Road mileage, 3,141; railway mileage, 567. Chief means of transport are 34 air lines. Telephone, telegraph and radio link Alaska with Canada and the United States. Education: 11,000 pupils in 108 public schools; 5,000 in schools maintained by U. S. Department of the Interior. Population of chief towns: Juneau (capital), 5,818; Anchorage, 11,060.

Hawaii consists of a group of islands, 8 inhabited, in the North Pacific Ocean, about 2,090 miles southwest of San Francisco. Total area, 6,454 square miles; population (1950), 499,794. Governor, secretary and judges are appointed by the president of the United States. Legislature of 2 houses—Senate of 15 members and the House of Representatives of 30 members. Agriculture is the chief industry. Sugar and pineapples take up 96% of cultivated area; coffee, animal products and bananas are also important. Chief exports: sugar, pineapples, coffee, bananas, hides and honey. Tourist trade is of considerable volume. Imports: manufactured goods, foodstuffs, oils, rice, lumber and fertilizers. International and island-to-island shipping is important; railway mileage, 25; motor vehicles, 128,000. Aviation: 9 airlines for trans-oceanic travel, 6 for inter-island. Telephone, radio-telephone, wireless and cable systems are in operation. Education: 118,000 pupils in 406 public and private schools and nearly 6,000 students at the University of Hawaii. Population of chief towns: Honolulu (capital), 245,612; Hilo, 27,019.

THE CITIES OF THE UNITED STATES

Its Centers of Culture and Industry

When George Washington became President of the United States, the population of the nation was almost entirely rural. The only towns of any considerable size were New York, Philadelphia, Boston and Baltimore, and the largest of these, New York, had only 33,131 people. As the country's population grew and its industry and transportation facilities expanded, there was an irresistible movement from rural areas to the towns, and many of the towns became the great cities of today. Signs in recent years indicate that a reverse trend is now in progress, from city to suburbs or country.

THE cities of the United States differ in many ways from those in other parts of the world. Some European cities have buildings that were old before Columbus was born, before America was even dreamed of. In the whole of the United States there are only a few city buildings built before 1700. Many Old World cities cling to narrow, winding streets that are just wide enough for one-way traffic to pass, almost brushing the walls. In many cases larger cars cannot go through. Most American cities have wide thoroughfares that take comfortably four lanes of automobiles. The downtown buildings in American cities are often very tall, with steel and concrete frameworks that enable them to reach toward the sky. In the rest of the world tall buildings are not typical, for structures of brick or stone cannot extend over seven or eight stories. Skyscrapers are American in origin and although many have now been built in other countries, they remain a symbol of the American city. The streets between such towering buildings often wend their way in shadow except during those few hours of each day when the sun is standing nearly overhead.

Just what is a city? Certainly it is a place where many people live together in a comparatively small area. But how

FAIRCHILD AERIAL SURVEYS INC.

LIKE A CITY IN A DREAM—THE LOWER END OF MANHATTAN ISLAND

Often the tops of the skyscrapers are veiled in mist so that the first view of New York City as one sails into the harbor seems unreal—so fabulous that it will soon vanish.

NEW YORK CENTRAL SYSTEM

THE BIRTHPLACE OF PAUL REVERE
The well-preserved house, with a steep, shingled roof, was built in Boston around 1660.

large such a place must be before it can be called a city is still a question.

There are about 43 cities in the world with populations of more than 1,000,000. Of these, 5 are in the United States. About 12,000,000 people live in Greater New York City alone! Almost one-fifth of the people of the United States live in some 150 cities of over 100,000. In this chapter we discuss only the country's larger or more important communities.

The number and great size of cities throughout the world is a relatively recent thing. Although there have been great cities for a very long time—even as long ago as the days of ancient Egypt and Mesopotamia—it is only within the last two hundred years that the world has seen their amazing growth. Cities today are changing, like boys who finally stop growing up and start filling out.

The modern American city depends upon vast transportation facilities for its existence. Food for the millions of people in and around Chicago, for instance, is not grown by the people of Chicago themselves. Rather, it is transported from farms in Kansas, Ohio, Florida or California. Since city people do not grow their own food, they must make something of use or value which they can sell or exchange for food. This explains the reason why the large modern manufacturing city came into being.

There are many kinds of cities. Some of the best known cities of the United States serve a governmental purpose. Washington, D. C., is the best example of this type, although most state capitals have government as one of their primary operations. Most cities are commercial and manufacturing communities. If we trace back to the origins of such cities as Boston and New York, we discover that they developed as port cities because they had protected harbors. Manufacturing and larger commercial activities came later. Some cities, such as Detroit, developed as manufacturing cities almost from the beginning. Others, such as Atlanta, have their origins in strategic locations with respect to routes of transportation. Cities often are born at the mouths of mountain passes and where rivers can be easily crossed.

The primary purpose of some cities is

EWING GALLOWAY

MEMORIAL AT NEW BEDFORD
Immortalizing the old whalers and their slogan, "A dead whale or a stove boat."

VIRGINIA STATE CHAMBER OF COMMERCE

STATE CAPITOL AT RICHMOND, DESIGNED BY THOMAS JEFFERSON

The central part of the Capitol was completed in 1792, and the wings added in 1905. The figure in Capitol Square in the foreground is of Washington, surrounded by six Virginia statesmen.

DALLAS CHAMBER OF COMMERCE

MEETING PLACE OF CHAMPIONS—THE COTTON BOWL IN DALLAS

The mammoth stadium, accommodating 75,000 people, is the site of the annual southwestern football classic. Its name is a reminder of the importance of Dallas in world cotton markets.

GENERAL MOTORS

THE HOUSE THAT GLASS BUILT—BOTH FUNCTIONAL AND BEAUTIFUL
Designed to admit a maximum of light, the glass-walled building is not only beautiful, but also ideally suited to research. It is part of the General Motors Technical Center near Detroit.

NEW YORK CENTRAL SYSTEM

MECCA FOR MERCHANTS—THE MERCHANDISE MART IN CHICAGO
The famous twenty-four-story building, two blocks long and one block wide, employs 26,000 people. Here thousands of manufacturers and wholesalers can display their "lines" for buyers.

recreation. Miami, Florida, grew in this fashion, as did Palm Springs, California. Some of the important cities in each state are cultural centers and are the sites of universities. For this reason such communities as Cambridge, Massachusetts, exert an influence much out of proportion to their size. Finally, there are military or defense communities. In Europe, many cities originated in particular places because their locations were readily defensible. In the United States, San Diego, California, is such a city; it has grown big because a military and naval base is located there.

In certain ways all large American cities are alike in their interior arrangement. Imagine a Typical City from which all other cities differ only in their details. In Typical City there is a central business area, in the center of which is a large park. Facing the park is the city hall. Wide avenues extend out from the main business section, and spaced along the avenues are lesser business centers. A river runs through Typical City. Along its banks, not far from the business center, is the industrial section. A dam on the river supplies factories with the necessary electrical power. Railroad lines haul heavy freight to and from the factories. Great wholesale warehouses store all the things the city people buy in the retail stores of the business centers. This railroad-manufacturing-wholesale section is the noisy, bustling, throbbing heart that enables the city to live. It is usually ugly, and often visitors to Typical City avoid this section where heavy trucks jam the streets, factories clatter and sooty smoke thickens the air. Yet if all this were to end, city bankers, grocers, bakers and others would soon find that they were unable to work or sell their goods.

Between the broad avenues of Typical City there are many blocks of homes. Those next to the noise and dirt of industry are rented cheaply. In this section closely packed apartment buildings house many families each. Farther away from

CALIFORNIANS INC.

LONGEST IN THE WORLD—SAN FRANCISCO–OAKLAND BAY BRIDGE

Eight-and-a-half miles long—half of it over water—the bridge passes through a tunnel at Yerba Buena Island in mid-bay. The view here is of the west half, from the island to San Francisco.

industry and business there are smaller individual houses. Many of these are owned by the families living in them. Still farther out from the business centers are mansions and large estates. Beautiful apartment projects and very large homes occupy a portion of the river bank. Scattered through the residential section are churches, schools, playgrounds and parks. On a bluff overlooking the river stands the state university with its broad, green campus. Typical City is imaginary but its features and arrangement are very like those of most American cities.

Look at a map showing all of the cities of the United States having over 100,000 people. You will notice at once that they form a pattern. About one-third are located within or near a triangle having its corners at Chicago, Boston and Richmond, Virginia. This great urban area has available a combination of raw materials, power, markets and labor necessary for big industrial development. Except for the Pacific coast cities, the western half of the country has few large cities. The dry, grassy plains, the deserts and the Rocky Mountains are not conducive to big-city development.

During the twentieth century, the growth of cities in the United States has been remarkable and swift. Some have grown 50 per cent in the last ten years or so. The most rapid growth is seen in the South and along the Pacific coast. Cities lying in the urban triangle have also for the most part grown, but at a relatively slow rate. Few cities in the nation have lost population in the past ten years, but their expansion is taking an unusual form. Because of the increased use of the automobile, the central area of large cities is gradually being abandoned to business and people are moving their residence to the suburbs. Throughout the nation small towns adjacent to cities are gaining population more rapidly than the cities themselves. It is these small but growing towns that will some day help to swell the ranks of the great cities.

LOS ANGELES CHAMBER OF COMMERCE

CALIFORNIA'S ROOSEVELT HIGHWAY SKIRTING THE BLUE PACIFIC

The beautiful coastal highway passes through the popular beach resorts in Los Angeles County, paralleling many stretches of fine, sandy beach such as the Santa Monica section shown here.

IN BYGONE DAYS AUGUSTA, MAINE, WAS A TINY INDIAN VILLAGE
Now an industrial city, the capital of the Pine Tree State, it reflects a modern skyline in the Kennebec River. The products of its mills include lumber, paper, textiles and shoes.

CITIES OF THE NORTHEASTERN STATES

Although there are many communities of importance and renown in Maine, New Hampshire and Vermont, these states of northern New England have no really large cities. Maine's capital is Augusta, but its largest city is Portland. A fine harbor makes Portland a thriving fishing center. Bangor, named for a hymn tune, is a transportation city in the northern part of the state. Bar Harbor is one of many famous vacation resorts along the coast.

Concord, of Revolutionary War fame and the capital of New Hampshire, is located on the Merrimac River. Nashua and Manchester, also on the Merrimac, use power from that river in the manufacture of textiles and leather goods. At the mouth of the Piscataqua River lies the important port and naval yard of Portsmouth.

Burlington is the chief city of Vermont and a popular resort center. The city overlooks beautiful Lake Champlain and is noted for its fine parks. At Barre, Rutland and Proctor there are important granite and marble quarries. Montpelier is the state capital.

In contrast to northern New England, Massachusetts, Connecticut and Rhode Island include many large cities of industrial and commercial importance. Boston, the largest of these and the capital of Massachusetts, is noted both for its present industry and for its past history. Its name, like those of many New England towns, was drawn from old England, the home of the original settlers on Massachusetts Bay. Each year the city's historic landmarks attract thousands of visitors. Boston has long been a cultural leader among American cities, and its churches, museums and literary shrines are famous. Here lived such outstanding figures as Paul Revere, Longfellow, Louisa Alcott and many others who played prominent parts in the nation's history and cultural life. The bustling commerce of the modern city's port draws ships from all over the world, and Boston's fishing industry is a major activity. Cambridge, across the Charles River from Boston, is the site of Harvard University, Radcliffe College and Massachusetts Institute of Technology.

The industry of Worcester, the second largest city in Massachusetts, is greatly diversified; here steel wire, grinding wheels and looms are manufactured. This city is the home of Holy Cross and Clark universities. In southern Massachusetts, Fall River and New Bedford have developed as textile manufacturing centers. Metal and electrical goods are made in Springfield, which stands on the banks of

A PROUD OLD PORTLAND MANSION
The Sweat Memorial in Portland, Maine, has a collection of paintings and fine art objects.

FACTORIES AND SMOKE STACKS BESIDE THE CONNECTICUT RIVER
Springfield, Massachusetts, which in colonial days had only a few gristmills and sawmills, today turns out a long list of manufactured necessities from fine paper to heavy machinery.

COMMERCIAL SECTION OF VERMONT'S LARGEST CITY: BURLINGTON

The city is a port on the eastern shore of Lake Champlain whose southern end is connected with the Hudson River by the Champlain Canal. The University of Vermont is on the outskirts.

NEW HAMPSHIRE'S STATEHOUSE, MADE OF LOCAL WHITE GRANITE

The imposing statehouse in Concord is guarded by statues of Franklin Pierce, Daniel Webster, General John Stark and John P. Hale—all history-making sons of New Hampshire.

the Connecticut River in the midst of a rich agricultural region.

Providence, named by Roger Williams, is Rhode Island's capital and only large city. Providence is a producer of jewelry and silverware. Newport, south of the capital, is a wealthy residential and resort community.

Connecticut has three large cities. Hartford, the capital and the largest city, is a major center for insurance businesses. New Haven and Bridgeport are ports and, together with Hartford, are manufacturers of metal products. Yale University, the third oldest institution of higher learning in the United States, is situated at New Haven. Waterbury, a somewhat smaller city, is a leader in watch production and has a thriving brass industry.

New York City is so much larger than the other cities of New York State, indeed of the whole nation, that it deserves special mention. To people of other lands, it is the model of all American cities, and it is truly an amazing development. Over 7,000,000 people live within the city limits. The population of the entire urban

MONKMEYER

TOWN HOUSES ON BEACON STREET, IN BOSTON'S BACK BAY
The Back Bay is the city's most exclusive residential district and one of its most charming thoroughfares is Beacon Street. It breathes an atmosphere of quiet, gracious dignity.

ROGER WILLIAMS LOOKS DOWN ON THE CITY THAT HE FOUNDED

The memorial is on a hill and the raised hand of the statue seems to be blessing the city of Providence. In the center is the steeple of the church that Williams established in 1638.

YACHTS AT ANCHOR IN THE HARBOR OF NEWPORT, RHODE ISLAND

Though Newport's days of glory as an ultrafashionable resort have declined somewhat, it is still a center of yachting. Many races—to Bermuda, Annapolis—start from here.

concentration numbers over 12,000,000. More than one-fourth of these people are foreign-born, and so cosmopolitan is New York City that one may eat French, Italian, Chinese, Spanish, German and Hungarian food—to mention but a few—in Manhattan's restaurants.

Manhattan Island

That part of the city called Manhattan is an island lying at the mouth of the Hudson River. The present great city grew out of the small colony founded there in 1626 by Dutch settlers, who called it New Amsterdam. On Manhattan are located many world-famous landmarks, among them Times Square and Broadway, with their lights, and Wall Street, the center of America's banking interests. Central Park extends for fifty-one blocks down the center of the island, itself an island of open country in a sea of great buildings. Adjacent to the park is the Metropolitan Museum of Art and the American Museum of Natural History. The Public Library, some blocks south of the park, is one of the finest in the world. At the southern tip of Manhattan and up the western shore are long lines of docks where ships from all parts of the world load and unload their cargoes and passengers. Ships approaching the docks must first pass the gray-green Statue of Liberty which stands in the city's harbor, a symbol of freedom known the world over. Rail lines converging on New York handle vast quantities of goods coming from and going to the rest of the nation. The city is so large that there is a great variety of manufactures and businesses, including clothing, publishing and advertising.

Transportation in a Metropolis

Many people who have business in Manhattan live in residential areas adjacent to the island and ride to work each day. A system of bridges, tunnels, busses and subways makes possible the transportation of people and goods from one part of the city to another. From New York's two great airports flights leave many times a day for Europe, South America and all parts of the United States.

MIDTOWN LANDMARK OF NEW YORK
The tallest structure is the RCA Building. At in summer. In addition to the broadcasting fa-

CITY—THE STEEL AND STONE SKYSCRAPERS OF ROCKEFELLER CENTER
its foot is a sunken plaza, used as a skating rink in winter and as an open-air restaurant
cilities and countless offices, there are a theater, gardens, shops and some other restaurants.

THE ALBRIGHT ART GALLERY IN NEW YORK STATE'S SECOND CITY
Buffalo's imposing gallery copies the severely beautiful lines of classic Greek architecture, even to the caryatids—the draped female figures supporting entablatures—on either side.

WATER-FRONT GRAIN ELEVATORS
Buffalo is one of the chief Great Lakes ports, through which quantities of grain pass.

THE CITY HALL ON NIAGARA SQUARE
Niagara Square is the heart of Buffalo, marked by the civic edifice of thirty-two stories.

ON YALE UNIVERSITY CAMPUS, THIRD OLDEST IN AMERICA

A view of the old library building from Phelps Hall. The college was founded in 1701, and later named for Elihu Yale, who had given money to the school. In 1887 Yale became a university.

PHOTOS, EWING GALLOWAY

A VIEW OF HARTFORD FROM THE CONNECTICUT RIVER

Connecticut's capital and largest city, Hartford, is also a vital center of trade and manufacturing, and it enjoys a great volume of the world's insurance business.

PHILIP GENDREAU

THE PULASKI SKYWAY SOARS OVER JERSEY RIVERS AND MARSHES

The 4-lane highway, of steel and concrete, is 3½ miles long, between Jersey City and Newark. It crosses the Hackensack and Passaic rivers on bridges that are 145 feet above the water.

PENNSYLVANIA STATE DEPT. OF COMMERCE

PITTSBURGH'S GOLDEN TRIANGLE—BIRTHPLACE OF THE OHIO

At the tip of the Golden Triangle, the Allegheny and Monongahela rivers come together to form the mighty Ohio. The Golden Triangle, or Point, is the business section of the city.

Buffalo, the second city of New York State, is located just south of Niagara Falls and is noted for flour milling and iron and steel working. Iron ore and grain are brought here by barges through the Great Lakes. South of Lake Ontario and along the Mohawk and Hudson rivers are Rochester, Syracuse, Utica and Schenectady, cities that have a variety of industry. Cameras and optical equipment are produced at Rochester, metal articles of all sorts at Syracuse, knitted goods at Utica and electrical appliances at Schenectady. Albany, the state capital, is also a commercial and industrial center.

The Cities of New Jersey

New Jersey, for its small size, has a number of cities. With the exception of Trenton, the capital, in the west center of the state, all these cities form part of a large urbanized area of New York City or Philadelphia. Newark, Jersey City, Elizabeth and Paterson lie close together and just across the Hudson River from New York City. Of these, Newark, a commercial and industrial city, is the largest. Camden, an important producer of canned soups and phonographs, is across the Delaware River from Philadelphia. Although Paterson is famous for its silk mills and Trenton for its pottery works, the industry of New Jersey cities is highly diversified. Atlantic City on the coast— with five sandy beaches—is a favorite summer resort for the people of the New York, New Jersey and Pennsylvania area.

Home of the Liberty Bell

Philadelphia, Pennsylvania's "City of Brotherly Love," has long been a shrine of American liberty. Here, in time-mellowed Independence Hall, the Declaration of Independence was signed in 1776. In and about the city are many other buildings and sites of historic significance which are visited yearly by thousands. The city's checkerboard pattern of streets was laid out when Philadelphia was founded and it remains today as one of the earliest examples of city planning in America. Located along the important Delaware River waterway, Philadelphia

PHILIP GENDREAU

THE "CATHEDRAL OF LEARNING"
The University of Pittsburgh's main building is a towering skyscraper with Gothic lines.

MAIN ENTRANCE TO THE VAST PHILADELPHIA MUSEUM OF ART

The museum has an unusually fine location on a hill overlooking the Schuylkill River at one end of Fairmount Park. Leading up to this entrance is the broad, tree-lined Parkway.

SKYLINE AND WATER FRONT OF BALTIMORE FROM FEDERAL HILL

Baltimore, on the west shore of Chesapeake Bay, is an important eastern seaport. The Patapasco River gives it a water front forty-five miles in length, with excellent docking facilities.

THE UNITED STATES NAVAL ACADEMY AT ANNAPOLIS, MARYLAND

The aerial view above gives a glimpse of the picturesque setting, the spacious grounds and the impressive buildings, in French Renaissance style, of the splendidly equipped Naval Academy.

DU PONT CO.

WILMINGTON, CAPITAL OF THE DU PONT INDUSTRIAL EMPIRE

In the heart of the industrial city is the du Pont Building (left center) housing a hotel, theater and business offices. It is connected by a bridge with the Nemours Building (right center).

is a city of diversified industry, where shipbuilding and oil refining are major activities. Like Boston, Philadelphia has been one of the nation's cultural leaders for many years.

West of Philadelphia is Reading, a textile city, and Harrisburg, the state capital and an important railroad center for lines to the west. Scranton, on the Susquehanna River in northeastern Pennsylvania, is a leading anthracite-coal-mining city. In western Pennsylvania, where the Allegheny and Monongahela join to form the Ohio River, lies Pittsburgh, named for the English statesman William Pitt. It is the iron and steel capital of America. Great barges bring iron ore from the mines of Minnesota, through the Great Lakes, to ports along Lake Erie. From these points ore travels by rail to Pittsburgh to be made into steel. Coal to fire the roaring blast furnaces of the steel industry comes by rail from West Virginia.

Baltimore, on broad Chesapeake Bay, has one of the deepest and best harbors on the Atlantic seaboard and is Maryland's only large city. Characteristic of Baltimore are the many rows of red brick residences with white stone steps which line the streets. At nearby Sparrows Point are great steel mills and oil refineries. To the south, Annapolis, the state capital, is the site of the United States Naval Academy. St. John's College, for men, is also at Annapolis.

The home offices of many of the country's largest manufacturing corporations are situated in Wilmington, Delaware's only really large city. Dover, the capital, in the center of the state, has a charming colonial air. The statehouse, which dates partly from 1722, has been the capitol since 1777. Dover serves as a shipping point for a rich farming area.

CITIES OF THE SOUTHERN STATES

Richmond, the capital and largest city of Virginia and former capital of the Confederacy, is a beautiful city, with broad avenues, spacious parks and graceful monuments. Cigarette manufacture is its most important industry. Norfolk, located almost at the mouth of Chesapeake Bay, is the second city in size. Its harbor at Hampton Roads and its rail connections with the interior of the country have made it a major coaling and naval port. Nearby Newport News and Portsmouth are shipbuilding centers.

West Virginia has no large cities. Huntington is the largest but Charleston, somewhat smaller, is the state capital. Nearby coal deposits make metal industries important in both cities.

Charlotte in south-central North Carolina is the only large city in that state. However, widespread tobacco cultivation and manufacture make such cities as Raleigh, the state capital, and Durham significant. Greensboro is a textile city and Wilmington is a port for cotton goods and lumber. Asheville, in the Blue Ridge Mountains, is a delightful summer resort.

Charleston, a port city of South Carolina, is one of the old cities of the South. Its homes and flowers are magnificent and each spring tourists throng to its beautiful parks and gardens. Columbia, the state capital, is a commercial and industrial city.

Atlanta, capital of Georgia, is a primary railroad hub for the southeast. Lines converge here after running south along either side of the Blue Ridge Mountains. Destroyed during the Civil War and then rebuilt, it is now the site of automobile-assembly plants, cotton-goods factories and ceramic plants. Rail lines from north, south and west meet in Savannah, the major port city of Georgia; lumber, naval stores and cotton move from this

STANDARD OIL CO. (N. J.)

THE GOVERNOR'S MANSION IN CHARLESTON, WEST VIRGINIA

The residence is a gracious one, in a simple classic style, surrounded by lawns and gardens. To the right is the dome of the state capitol, one of the most beautiful in the nation.

NORTH CAROLINA NEWS BUREAU

NOT A ROLLER COASTER, BUT THE COLISEUM AT RALEIGH

The futuristic structure is an indoor arena in North Carolina's capital. It holds ten thousand persons and consists of concrete parabolas apparently supported only by green glass walls.

CHARLOTTE CHAMBER OF COMMERCE

BUSINESS DISTRICT OF CHARLOTTE

The chief city of North Carolina is also the principal industrial center of both Carolinas.

port and from Brunswick farther south. Savannah's winding streets, small parks and pre-Civil War buildings make it an extremely picturesque city.

City growth in Florida has been particularly rapid in the past quarter-century. Jacksonville, at the mouth of the St. John's River, is the Atlantic port and rail gateway to the state. The lower east coast, with Miami as the major city, consists of a long series of resort towns where the winter warmth is in sharp contrast to the cold of northern states. Tampa, on the Gulf of Mexico, is a port for trade with countries of the Caribbean. Its population, partly of Spanish descent, is engaged in cigar-making and shipbuilding. St. Augustine, the oldest continually inhabited city in the United States, was founded in Florida by the Spanish in the year 1565. Tallahassee is the state capital. Pensacola is an important naval air base.

In Birmingham, Alabama, the iron and steel center of the South, huge blast furnaces glow red against the night sky. They are made possible by nearby sources of coal and iron ore. Montgomery, an older city than Birmingham, was once the capital of the Confederacy and today is the state capital. Located near the agriculturally important Black Belt of Alabama, it has been a commercial city of some importance. Mobile was second only to New Orleans as a southern port in 1850. After relative inactivity at the turn of the century, Mobile is now resuming its growth. Harbor deepening, coal traffic by barge from the interior and natural gas and oil supplied by pipeline combine to make Mobile attractive for such industries as shipbuilding, paper mills and aluminum manufacture.

Growth of Cities in Mississippi

It is only within about the last forty years that Jackson, capital and largest city in Mississippi, has shown rapid growth. Light industries of various sorts are characteristic of the city. Meridian is a commercial city serving the agricultural region around it. Gulfport, Biloxi and Pascagoula are beach resorts and fishing centers on the Gulf coast.

Of Tennessee's four large cities, two are in the eastern part of the state along the banks of the Tennessee River. Chattanooga, the larger, derives hydroelectric power from nearby dams. This power, together with neighboring coal and iron deposits, enables the city to produce a variety of metal goods. Knoxville is a center for marble quarrying, wood products and, most important, cotton textiles. The state capital, Nashville, located in a lush agricultural area, is the site of several colleges and Vanderbilt University. Tennessee's western boundary is the Mississippi River, and on its high bluffs is Memphis, the largest city in the state. Memphis has shipped cotton and timber up and down "Ol' Man River" since the days of paddle-wheel steamers.

The Ohio River flows along the northern boundary of Kentucky and on its banks is Louisville. Like Memphis, Louisville has been an important river port since its founding by George Rogers Clark in 1779. Together with Lexington, near the center of the state, Louisville is

A RAILWAY INTO THE CLOUDS

The cable railway that ascends Lookout Mountain in Chattanooga is the steepest in the world.

JACKSON'S STATUE IN NASHVILLE

The bronze statue of the famous President is near his beautiful old home, The Hermitage.

OLD CHARLESTON SLAVE MARKET
The building was the source of house and field labor for a large area in South Carolina.

CHARLESTON'S PINK HOUSE
Among the attractions for visitors in the historic city is this pre-Revolutionary tavern.

PHOTOS, BLACK STAR

PALMETTOS AND LIVE OAKS ALONG THE BATTERY IN CHARLESTON
The popular walk overlooks Charleston Harbor and the historic fortifications at Fort Sumter. Near the promenade is a small park containing monuments and mementos of the city's history.

JALOUSIED WINDOWS AND BALCONIES ON A SAVANNAH STREET
The old houses in Savannah all are built somewhat alike. Most of them rise three or four stories above a basement. From the street one must climb steep stoops with iron railings.

BIRMINGHAM CHAMBER OF COMMERCE

NIGHT SHIFT IN BIRMINGHAM
When the blast furnaces are run at night, the dramatic glare may be seen for miles away.

SAWDERS FROM CUSHING

NEW CITY OF THE OLD SOUTH
Atlanta, built on the ashes of the ante-bellum city, has become an energetic industrial center.

BLACK STAR

MEMORIAL TO SOLDIERS AND SAILORS IN MOBILE, ALABAMA
Mobile is a city of beautiful parks. There are thirteen public ones, including Memorial Park (above) and famous Bienville Square, as well as several private parks open to visitors.

EWING GALLOWAY

THE STATELY CAPITOL BUILDING AT MONTGOMERY, ALABAMA
One of the most magnificent structures in Montgomery is the classic domed capitol building with its portico of fluted columns, set high on beautifully landscaped grounds.

LITTLE ROCK CHAMBER OF COMMERCE

THE TERRITORIAL CAPITOL IN LITTLE ROCK, ARKANSAS
The white clapboard building was the capitol before Arkansas became a state. Surrounding it are private homes that have authentic furnishings dating from the territorial period.

CITY OF MIAMI NEWS BUREAU

YACHT BASIN IN MIAMI CITY WITH ITS CHARTER FISHING FLEET

Catering to the vacationer is the major industry of most of Florida. Miami, the glamour city adds to its list of attractions one of the finest fleets of charter fishing vessels in the world.

FLORIDA STATE NEWS BUREAU

TUNEIVORN HOUSE IN ST. AUGUSTINE, FLORIDA'S OLDEST CITY

Today the building contains relics of St. Augustine's history since its settlement by the Spaniards in 1565. The surrey (right) for leisurely sightseeing is in keeping with the city's mellow charm.

VIEW OF JACKSONVILLE, FLORIDA, FROM MAIN STREET BRIDGE

Jacksonville, one of Florida's largest cities, is an industrial and commercial center as its skyline would indicate. It is an important world port as well as being the principal gateway to Florida.

MUNICIPAL PLAZA IN BEAUTIFUL CITY OF JACKSON, MISSISSIPPI

Across the Municipal Plaza, landscaped with colorful, formal flower beds, is the Hinds County Court House. At the left is the City Hall that was built in 1864 during the Civil War era.

noted for the fine horses that are bred in the surrounding Blue Grass region. Frankfort is the state capital.

Louisiana has one major city, New Orleans, whose population is concentrated along the banks of the Mississippi and adjacent Lake Pontchartrain. Great wharves and warehouses line the city's harbor, and from here grain, cotton, petroleum and machinery are shipped to ports throughout the world.

New Orleans was founded over two hundred years ago by the French and many of its people are of French descent. No visitor should leave New Orleans without seeing the buildings in the French Quarter, where balconies of intricate iron grillwork lend an Old World beauty to

CHARTRES STREET IN THE ROMANTIC VIEUX CARRE OF NEW ORLEANS
The Vieux Carré, the old French quarter of the city, recalls the days when it was the capital of the French colony. Along the narrow streets there are rows of lacy grillwork balconies.

STANDARD OIL CO. (N. J.)

THE WIDE DAM IN THE OHIO RIVER AT LOUISVILLE, KENTUCKY
River traffic moves around the dam and nearby falls by means of a canal and locks. Overhead is the George Rogers Clark Memorial Bridge, connecting with Jeffersonville, Indiana.

TOWBOAT AND BARGES ON THE MISSISSIPPI OFF BATON ROUGE
Soaring up from the center of the city is the skyscraper capitol building. Besides being the home of the state government, Baton Rouge is a busy river port with modern docks.

A STATUE OF A COWBOY OVERLOOKS OIL FIELDS IN OKLAHOMA CITY
Since oil was discovered in 1928, Oklahoma City has become a forest of derricks. More and more wells have been sunk until today they are within a few feet of the capitol building itself.

mellowed houses and gracious hotels.

Up river from New Orleans is Baton Rouge, the state capital and an oil-refining center. The capitol building is a towering skyscraper. Like Baton Rouge, Shreveport, on the Red River, is important because of its oil refineries.

Near the geographical center of Arkansas and on the Arkansas River is Little Rock, the capital and largest city in the state, as well as its commercial crossroads. Here goods from east and west meet for exchange. Farther up the Arkansas River is Fort Smith, a secondary commercial city. Southwest of Little Rock is Hot Springs, a health and pleasure resort adjoining Hot Springs National Park.

Oklahoma City, Oklahoma, was formerly a small community in an Indian territory. Then oil was discovered there, and the town suddenly began to grow in size and wealth. Today oil pervades its atmosphere, and great derricks used in drilling have covered nearly all of the city's available space, even invading the grounds of the state capitol building. In Tulsa, once a sleepy Indian town, the oil refineries vie with modern skyscrapers.

Texas is so immense that there is room for many cities. Houston, the largest of these, is the biggest port on the Gulf coast. Once it was an inland town, but the ship canal dug to Galveston Bay has provided it with a harbor. The port city of Galveston is on an island paralleling the coast and is connected with the mainland by causeways. San Antonio is the site of the historic old mission, the Alamo, and several military bases. Manufacturing and trade across the Mexican border are its chief occupations. Dallas, a major banking and insurance center, is also a rail junction point and a great cotton market. In addition to its cotton mills, nearby Fort Worth has large stockyards and meat-packing enterprises. El Paso, in the western corner of Texas, is located in the fertile irrigated valley of the Rio Grande. The many ranches in the area supply its stockyards with beef and cattle. El Paso, across the border from Mexico, has a Spanish atmosphere. Corpus Christi is the port for southwest Texas. Austin, near the center of Texas, is the capital.

BLACK STAR

CRADLE OF TEXAN LIBERTY
The Alamo (right) in San Antonio, scene of a tragic siege in the Texas-Mexico struggle.

FOLEYS—KING-SIZE DEPARTMENT STORE IN HOUSTON, TEXAS
A major attraction for shoppers is the mammoth emporium occupying a square block in Houston. There are few windows above the street floor, which permits more display space inside.

RAYMOND LOEWY ASSOCIATES

MUSEUM OF SCIENCE AND INDUSTRY IN CHICAGO'S JACKSON PARK
A domed rotunda and the Ionic columns of portico and colonnade rise impressively above Columbia Basin. The museum contains models and exhibits of industry's greatest inventions.

CITIES OF THE NORTH CENTRAL STATES

Ohio's many large industrial cities are grouped along Lake Erie, in the central part of the state, and along the Ohio River. The largest is Cleveland, the state's principal lake port and the home of a thriving steel industry. Iron ore is brought here by Great Lakes barges to feed Cleveland's furnaces. Some of it is shipped on by rail to the nearby industrial cities of Youngstown and Canton. Toledo, Ohio's second great lake port, is also a principal rail center, with lines extending in all directions. Akron, one of the world's largest rubber-manufacturing cities, is a major producer of tires and tubes for trucks and automobiles. Columbus, in central Ohio, is the capital and is industrially important for its clay products. United States Government testing laboratories are located in Dayton, which is a city of diversified industry that includes the manufacture of office equipment and refrigerators.

Cincinnati, on a northward bend of the Ohio River, has a history dating from the Revolutionary War. Set in the prosperous Ohio Valley farming area, it has grown from a frontier fort into a great rail and commercial center. One of the largest soap-manufacturing plants in the world is located here.

Indianapolis, in Indiana's rich corn belt, is the capital and largest city in the state, with important flour mills and meat-packing plants. Fort Wayne and South Bend, in the northern part of the state, are

THE LOOP—A HIVE OF COMMERCE IN STEEL, CONCRETE AND STONE
A maze of surface transportation—broad auto expressways and streets, bridges that open for boats on the Chicago River, railroads everywhere—forms the apron of Chicago's Loop.

THE CITIES OF THE UNITED STATES

AUDITORIUM IN BATTLE CREEK
The auditorium in the breakfast-food city is the gift of the cereal-maker W. K. Kellogg.

LINCOLN, IN THE CITY HE LOVED
Andrew O'Connor's statue of Lincoln before the capitol building in Springfield, Illinois.

known for their flour production and the manufacture of farm machinery, sewing machines and clothing. Gary, on the south shore of Lake Michigan, is primarily a steel-producing city.

Illinois is dominated by one gigantic city, Chicago. Spreading out along the southwest shore of Lake Michigan, it is second only to New York in size, with over 3,500,000 people dwelling within its city limits. Over 5,500,000 more live in the metropolitan area of Greater Chicago. The meeting here of the industrial East and the agricultural West has made it possible for a lakeside village to develop into a leading metropolis of the world within a single century. Today Chicago is the major rail hub of the Midwest and from its heart a complex maze of lines reaches out in all directions like the spokes of a wheel. Cattle and hogs from the nearby corn belt fill its stockyards and make it the principal meat-packing city of the nation. Iron ore and grain brought in by lake barges supply its steel and flour-milling industries. For many years it has been a big manufacturer of farm machinery. Between the massed skyscrapers of

THE EDISON INSTITUTE

HENRY FORD'S LASTING TRIBUTE

A replica of Independence Hall is the entrance to Edison Institute in Dearborn, Michigan.

its business section and the lake shore is a long park containing museums and art galleries. Here the people come on hot summer days to be cooled by lake breezes. However, these same breezes blow cold and strong in winter and have given Chicago its nickname of the "Windy City."

Springfield, the state capital in central Illinois, was the home of Abraham Lincoln and is the site of his tomb. Each year it is visited by many who come to pay their respects to the great leader. Peoria, between Springfield and Chicago, is situated in a rich agricultural region and is the center of an enormous grain trade.

Although it is noted chiefly for its automobile manufacture, Detroit, Michigan, is also a port city. Founded by the French on the Detroit River, it has water connections with Lakes Erie and Huron and is the largest city in the state. Nearby Flint is also an automobile-factory city. Grand Rapids, on Lake Michigan, has been nicknamed the "Furniture Capital" of the nation. Kalamazoo, widely known because of its odd name, is the site of an important paper industry. Lansing is Michigan's state capital.

BLACK STAR

TO THE BRAVEST OF THE BRAVE

On the capitol grounds in Des Moines is a heroic column to Iowa's Civil War veterans.

TOWERS ON THE DETROIT RIVER—THE CITY THAT CARS BUILT

The buildings of downtown Detroit turn inward from the Detroit River, the busy strait between Lake St. Clair and Lake Erie. The names of Detroit's founder, Antoine de la Mothe Cadillac, and of the men who gave impetus to its growth at the start of the automotive age—Ford, Buick, Dodge, Olds, Chrysler—are known in every corner of the motoring world.

PLANE'S-EYE VIEW OF COLUMBUS, OHIO

Columbus, capital of Ohio, was named for Christopher Columbus. The town was laid out in 1812 and became a city in 1834. The state capitol building is of local limestone, built in Grecian style. Columbus is the site of Ohio State and Capital universities, and of Starling Ohio Medical School. Fourteen railroads converge in this city at the Union Station.

THE UNIVERSITY OF CINCINNATI

The Student Union Building represents a graceful combination of modern building methods and traditional styles of architecture. The campus is in picturesque Burnet Woods Park.

CHARLES PHELPS CUSHING

THE BUSY, BACKYARD CUYAHOGA

A tipple loads a lake ore boat with coal for Detroit in Cuyahoga River Flats—the cluster of blackened mills, factories and rail yards a stone's throw from downtown Cleveland.

CINCINNATI CHAMBER OF COMMERCE

CINCINNATI'S IMPOSING UNION TERMINAL

This handsome building is one of the most modern railroad stations in the world. The huge clock is a landmark. The city is in an area of heavy population, and this has helped to make it a leading railroad center. More than a dozen eastern and southern railroad lines enter Union Terminal. The city also owns and operates the Cincinnati Southern Railway, to Chattanooga, Tennessee.

EWING GALLOWAY

HEART OF INDIANAPOLIS—THE SOLDIERS' AND SAILORS' MONUMENT

The great shaft, 284 feet high, is in the center of the city and commemorates Hoosiers who have given their lives in service to their country. At the top is a gigantic statue of Victory, 38 feet high. On the eastern and western sides of the base, water cascades down into pools. Looking up the broad avenue from Monument Place, you can see the handsome domed Capitol.

THE CITIES OF THE UNITED STATES

Milwaukee is by far the largest city in Wisconsin and, like Chicago, it is a port on Lake Michigan. It manufactures heavy machinery and has many large flour mills and breweries. Madison, the state capital and seat of the state university, is a pleasant city located amid lakes and rolling hills.

Minneapolis and its twin city St. Paul lie on either side of the Mississippi River in Minnesota. The former is noted for its great flour mills. St. Paul, the state capital, is a rail center and an important meat-packing city. There are many ponds and lakes about the "Twin Cities" that add a fresh and attractive air to the bustle of metropolitan life. In northern Minnesota, at the extreme western end of Lake Superior, stands Duluth, through which flows iron ore from the nearby Mesabi mines to feed the country's blast furnaces. Wheat from the Dakotas and lumber from Minnesota also come to Duluth for shipment by lake barges to the states farther east.

Near the center of Iowa is the state capital and largest city, Des Moines, a

A BIT OF VENICE IN INDIANA
The dome is the culminating point of the Fort Wayne Court House, which has a look of Venetian baroque. It is in the center of the city—a railroad hub and heart of a farming region.

THE WHITE CITY OF ST. PAUL ABOVE THE FATHER OF WATERS
A stern-wheeler excursion boat, riverside retaining walls, the piers of Robert Street Bridge and the massive buildings of St. Paul cast a glow on the Mississippi. St. Paul grew up as a banking and railroad center; its cross-river twin, Minneapolis, as a flour- and saw-milling town. Today they work together as the industrial capital of the northern states.

COFFMAN MEMORIAL UNION OF THE UNIVERSITY OF MINNESOTA

The forthright-looking building is a fitting tribute to a former president of the university. It is used as a residence for the many students who come from outside Minneapolis.

FOUNTAINS SPLASH IN FRONT OF UNION STATION IN ST. LOUIS

More than twenty railroads pass through the huge station, as the city is a crossroads for the whole country. The four-story building, of stone, is in a modified Romanesque style.

FISHING CONTEST FOR SMALL FRY IN KANSAS CITY, KANSAS
There are a number of lakes in the city which are open to young Izaak Waltons for one morning a week during the summer. This is Big Eleven Lake, kept stocked by the state fish hatchery.

ACRES OF STOCKYARDS ALONG THE KAW RIVER AT KANSAS CITY
Stockyards and packing houses are on both sides of the river, some in Kansas City, Missouri and some in Kansas City, Kansas. They can process thousands of livestock in a day.

MONUMENT IN TOPEKA   **WISCONSIN'S CAPITOL**
A pioneer mother guards her children, with a rifle across her knees, on the grounds of the Kansas state capitol. The handsome structure in Madison, Wisconsin, has a ribbed dome.

A DREDGE REMOVING SILT FROM THE MILWAUKEE RIVER CHANNEL

The river flows through Milwaukee and is a feeder for the city's sheltered harbor on Lake Michigan. In the background the Wisconsin Avenue Bridge, a movable one, is raised.

LAWNS BETWEEN COUNTY AND CITY BUILDINGS IN OMAHA

On the farther side of the street, at the corner, is the rather massive-looking City Hall. The smooth lawn, with star-shaped beds, is in front of the Douglas County Court House.

THE FRONTIERSMAN STATUE ON THE CAPITOL GROUNDS AT BISMARCK
In the distance is the capitol building of North Dakota. It is a structure in the modern style and seems all the taller for the open space and low buildings that surround it.

corn-belt community of meat-packing houses, creameries and farm-equipment factories. The Iowa State Fair, held in Des Moines each year, is one of the country's big agricultural events. Other important food-processing and commercial cities in Iowa are Sioux City, Council Bluffs, Cedar Rapids and Dubuque.

Two of the nation's great Midwestern cities lie in Missouri, St. Louis and Kansas City. St. Louis, near the junction of the Mississippi and Missouri rivers, is a great river port. It was founded by the French, and its industrial history goes far back to the days when trappers brought their furs down the Missouri to its market. Today it is a prominent flour-milling and meat-packing city. Kansas City, on the Missouri, has a twin, Kansas City, Kansas, just across the river. Both are food-processing cities. Jefferson City, in central Missouri, is the state capital.

The largest city in Kansas is Wichita, on the Arkansas River. Set in the winter-wheat belt, the city does a large milling business. The prosperous farms of the region provide a great and ready market for the farm supplies in which Wichita dealers specialize. The state capital, Topeka, is just up river from Kansas City.

Ever since the days of the covered wagons, Omaha, Nebraska, has been a key point in transcontinental traffic and today it is a railroad center for lines running west. It is also a great agricultural and livestock market, and its food-processing industry is large. The thriving commercial city of Lincoln is the capital of Nebraska. Its capitol building combines the skyscraper and classic forms in one of the country's more beautiful government buildings.

The Dakotas have none of the very large cities that generally develop from an industrial economy. Sioux Falls, South Dakota, is a grain city with many flour mills, and Pierre, almost in the exact middle of the state, is the capital. The tourist trade is drawn to Rapid City by the famous carvings of United States presidents on a mountainside in the nearby Black Hills. Fargo, on the Red River, is North Dakota's largest city and a commercial and wheat-handling center for the surrounding agricultural belt. The state capital is Bismarck.

CITIES OF THE WESTERN STATES

Transportation, mining or livestock have been largely responsible for the growth of cities in the less densely populated Rocky Mountain states. Although the region covered by these states is generally arid, local supplies of irrigation water have made intensive farming possible in some areas. In such sections, agricultural communities have developed.

Despite its vast area, Montana, the largest of the mountain states, has no very big cities. Butte, the largest, is a world-famous city of copper mines and smelters and was once nicknamed the "richest hill on earth." In order to ship its ores and metals, Butte has developed fine railroad facilities. Mining is also carried on at Great Falls and Missoula. Billings, on the Yellowstone River, is a trade center for southern Montana. Helena, the capital, is a distributing point for a large stock-raising region.

Wyoming is a state of spectacular scenery, where cattle and sheep herding are the most important activities. Cheyenne, the capital and largest city, is situated in the midst of good range land and is a shipping head for beef cattle and sheep to eastern markets. Casper, the next town in size, is a producer of petroleum. Cody, named for "Buffalo Bill" Cody, is the eastern gateway to Yellowstone National Park and is a stopping place for tourists.

Colorado is another state of scenic grandeur and variety and is somewhat more densely settled than Montana and Wyoming. Denver, "The Mile High City" and capital, lies just at the foot of the Rockies and commands a superb view of the mountains. Its cool summers attract many vacationers. The surrounding irrigated farmland and pastures make it a great market city and it also has some

CENTRAL AVENUE, ALBUQUERQUE, METROPOLIS OF NEW MEXICO
Mission and frontier Spanish architecture stand side by side with classical Greek and modern American on Central Avenue. Albuquerque is a health resort and business center of a rich area.

light industries. It is the home of the University of Denver and several other institutions of higher learning. The Denver Art Museum contains an outstanding collection of American Indian art. The city also has a fine civic symphony orchestra and other musical and art organizations. A United States mint is located in Denver. Colorado Springs, near the base of Pike's Peak, is a famous health resort. Deposits of iron and coal near Pueblo are responsible for that city's steel mills.

Santa Fe is the oldest city in New Mexico and the state capital. It was founded by the Spanish in 1605, and its picturesque pueblo-type architecture makes it an interesting place to visit. Before the coming of the railroads to the southwest, it was the terminus of the Santa Fe trail from Missouri. Today its elevation makes Santa Fe an important health resort. Al-

MONTANA CHAMBER OF COMMERCE

BUTTE, MONTANA—COPPER TOWN
Marcus Daly, copper-mining millionaire, looks over the city his enterprise helped to build.

buquerque, on the Rio Grande, is the commercial heart of the state and to it come the products of the nearby irrigated valley farms and ranges.

A bright green oasis in the desert is Phoenix, Arizona's capital and largest city, situated on the Gila River. Its rail connections make it a major distributing point for the southwest. Tucson, formerly capital of the state, is a commercial city near the copper-mining district of southeastern Arizona. In the northern part of the state, Flagstaff and Williams serve as gateways for tourist traffic to the Grand Canyon region.

Much of Utah's population is concentrated on the eastern shore of Great Salt Lake, where the Mormons founded Salt Lake City in 1847. Rich irrigated land that produces grain, sugar beets and other vegetables surrounds the city. Wide avenues and Mormon churches add to the attractiveness of the city. Ogden, to the north, is a rail center. Bingham is the site of one of the nation's biggest copper mines.

Idaho has no large cities. Boise, the state capital, is the major commercial city and lies at the heart of an irrigated farm area. Pocatello has great railroad shops,

SANTA FE RAILWAY

ADOBE DWELLING IN SANTA FE
The "Oldest House in U. S. A."—built perhaps before 1600—recalls an Indian settlement.

DENVER'S CITY AND COUNTY BUILDING AND THE GREEK THEATER
Columns of the Greek Theater frame the graceful tower of the City and County Building. Denver, the "mile-high city," has several colleges and a lively interest in the arts.

IDAHO STATE CHAMBER OF COMMERCE

AT BOISE—IDAHO'S MAJESTIC COPY OF THE NATIONAL CAPITOL
The dignified façade and the columns of the tower seem to hold high the central dome. Within are legislative chambers, government offices and the state historical museum.

stockyards, cement plants and food-processing facilities. Coeur d'Alene is both a mining center and resort town.

Nevada has two well-known cities, Las Vegas and Reno, both mining towns that are also amusement and tourist centers. The capital, Carson City, was once a Pony Express station. A hundred miles northwest of Las Vegas is Frenchman's Flat, where the Atomic Energy Commission uses five thousand square miles of wasteland as an atomic testing ground.

In contrast to the mountain states of the interior, the more populous Pacific coast states of California, Oregon and Washington have many large cities. Southernmost of these is San Diego, California, which developed from an old Spanish mission and is the oldest city in the state. In many ways its atmosphere reflects the peaceful and colorful spirit of mission days. Landlocked San Diego Bay is an excellent natural harbor. It is a naval base and the dry, warm climate makes year-round naval and air operations possible.

Farther north along the coast is Los Angeles, commercial and tourist center of southern California and the largest city west of Chicago. Almost 2,000,000 people live inside the city limits and the metropolitan area contains over 4,000,000. Founded in 1781 as a tiny Spanish settlement, it has spread out until today it embraces the greatest area of any city in the United States. It is the hub of the motion-picture industry and leads all other cities of the state in wealth, manufacturing, commerce, aviation and ocean shipping. The delightful, sunny climate has made it one of the nation's outstanding recreational spots. The University of Southern California and the southern branch of the University of California are also in Los Angeles.

The second great city of California is situated on San Francisco Bay, which is separated from the Pacific Ocean by the Golden Gate channel. San Francisco takes its name from a Spanish mission founded about 1776, but it did not begin to grow until the gold rush days of 1849. It had already become a prosperous community by 1906, when it was devastated by fires following an earthquake. Its citizens immediately began to rebuild their city, and today it is a major supply base and distribution center for Pacific coast

BRIGHAM YOUNG MONUMENT BASE	TO "THE MIRACLE OF THE GULLS"
The figure is on a Salt Lake City statue to the man who led the colonization of Utah.	Tribute in Temple Square to gulls that saved the pioneers from a grasshopper plague.

THE MORMON TABERNACLE AND TEMPLE IN SALT LAKE CITY
Six ornate spires of the Mormon Temple rise in stately watchfulness over the broad, fertile valley of Great Salt Lake. The rounded, white roof shelters the famous Mormon Tabernacle.

agricultural, manufacturing and mining industries. Its landlocked harbor is one of the finest in the world. The famous Golden Gate Bridge suspended across the outer mouth of the harbor has the longest span of any bridge in the world. It is really a series of bridges, with a total length, including its approaches, of eight and a quarter miles. Ocean liners can easily pass beneath it. So steep are many of the hills on which San Francisco is built that street cars run on cog rails.

The capital of California is Sacramento, set in the Great Valley which extends north-south through the center of the state. The fruits and vegetables that grow around the city are canned in its large food-processing plants. Stockton and Fresno are located in good agricultural areas. The former, once a gold rush city, now manufactures farm machinery. Fresno is known as a "raisin capital."

In Oregon most of the population is in the beautiful Willamette and Columbia

SAN XAVIER DEL BAC MISSION
The Indian mission church near Tucson is an example of Spanish baroque in America.

FREIGHT YARD OF THE UNION PACIFIC IN CHEYENNE, WYOMING
Cheyenne, a city that grew up as a main division point of the railroad westward, thrives today as the shipping and trade center for a region rich in the production of minerals and food.

ARIZONA'S STATEHOUSE AND THE CAPITOL PARK IN PHOENIX
In the park grounds before the state capitol, a memorial to the sons of Arizona who served in World War I stands in quiet dignity among the swaying palms and clipped evergreens.

A GUEST RANCH OF PHOENIX WITH AN AIR OF MOORISH SPLENDOR
At a New World retreat in an Old World, Moroccan key, vacationists bask beneath a bright desert sun. Phoenix is lively all through the winter with the comings and goings of tourists.

STATE LIBRARY AND SUPREME COURT BUILDINGS, CARSON CITY

Nevada's State Library and Court House stand in dignity where once was the surge of frontier life. Carson City, the smallest state capital in the United States, has had history enough for a town several times its size. Named for Kit Carson, intrepid scout and soldier, it was a center of supply during Indian wars and the great gold and silver booms.

PHOTOS, BLACK STAR

A DAYTIME VIEW OF VIRGINIA STREET, THE WHITE WAY OF RENO

Reno's slogan and the neon signs announcing night clubs and bingo parlors tell but a small part of the city's story; and its assembly-line divorce court, but a fraction of the remainder. The University of Nevada, the banks, the ski slopes nearby in the Sierra Nevada make it an educational, commercial and recreational center for the state and bordering counties.

ALL YEAR CLUB OF SOUTHERN CALIFORNIA

A BEACON IN NIGHT-TIME LOS ANGELES

The Los Angeles City Hall, dramatically lighted at night, is part of the city's beautiful Civic Center. Proud of its rapid growth and the beauties of its scenery and climate, Los Angeles is the fourth largest city in the United States. Much of its growth can be traced to the motion-picture industry, which chose Hollywood, part of Los Angeles, as its center.

THE CITIES OF THE UNITED STATES

IN BALBOA PARK, SAN DIEGO
A long way from Elizabethan London is San Diego's replica of Shakespeare's famous playhouse, the little Globe Theater.

valleys. Portland, built on both sides of the Willamette River near its confluence with the Columbia, is the state's major city. Since the Columbia River is large enough for sea-going vessels, Portland is an important west coast port. Farm and forest products are among the principal industries, and the city's wool textile mills employ many people. Because of its beautiful location, Portland is sometimes called the "City of Roses." Salem, the capital, is an agricultural, textile and food processing city.

Seattle, on Puget Sound, is Washington's largest city and an important port for trade with Asia and Alaska. Tacoma's lumber industry is one of the largest in the country, and paper mills, copper smelters and electrochemical plants add to the city's industrial might. Spokane, in the fertile wheat area east of the Cascade Mountains, is a leading commercial and rail center.

We have looked at all the large and most important cities of the United States. However, the list of them will not remain

PHOTOS, BLACK STAR

WHERE VINE CROSSES SUNSET IN THE TOWN THEY CALL COLOSSAL
West coast studios of NBC stand at one of Hollywood's busiest intersections. In and around the city, actually a part of Los Angeles, are other stations and many motion-picture studios.

PHOTOS, PHILIP GENDREAU

FISHERMEN'S WHARF
Masts and a tangle of rigging lace San Francisco's water front where the fishing fleets put in.

SAN FRANCISCO, FROM NOB HILL
Skyscrapers stand out against San Francisco Bay and hide the approach to Oakland Bridge.

CBS

HOLLYWOOD'S MODERN TELEVISION CITY, FOUR STUDIOS IN ONE
The Columbia Broadcasting System's home on the west coast, Television City, brings together all the various CBS departments of TV production under one big roof.

NORTHERN PACIFIC RY.

PORTLAND—SHIPPER OF WOOD AND WOOL, PAPER AND GRAIN
On the west bank of the Willamette River, near its union with the Columbia, stand the skyscrapers of Portland—trading headquarters for the industries of the Northwest.

BLACK STAR

STATE CAPITOL IN SALEM—A MONUMENT TO THE PIONEER WEST
In white marble, of modern design, Oregon's capitol serves the needs of government and pays tribute in sculpture and murals to the hardy men and women who helped to build the state.

THE CITIES OF THE UNITED STATES

the same over a long period. Within a few years many towns and smaller cities not mentioned here will have grown large.

Now that you have seen the large communities in some detail, imagine yourself in an airplane flying high over the land. Just below is the Mississippi River and St. Louis. Off toward the setting sun, beyond the Rocky Mountains, is the Pacific. To the east, in dark shadow, sprawl the Appalachian Mountains, and past them roll the waters of the Atlantic. Here and there, like tiny ant hills scattered through the patchwork of forests, fields and desert land, lie the great cities. You can see how few miles out of the total they occupy. Yet their seeming tininess belies their importance, for in them stirs the nation's industrial power.

BY JOHN R. DUNKLE

FLOATING BRIDGE AT SEATTLE
The longest floating bridge in the world steps across Lake Washington on concrete pontoons.

SEATTLE'S MODERN PIERS AND THE WATER FRONT ON ELLIOTT BAY
Freighters, tankers and liners dock at Seattle, metropolis of the Northwest, beautifully situated on an isthmus between the Pacific inlet, Puget Sound, and fresh-water Lake Washington.

© Hileman

GOING-TO-THE-SUN MOUNTAIN, IN GLACIER NATIONAL PARK, MONTANA

This peak can be reached by a trans-mountain highway from the eastern boundary. The wild region of this northernmost park of the United States was once the hunting ground of the Blackfoot Indians. Ages before that, the region was covered by a prehistoric sea and its sediments have left horizontal streaks of colored rock in the mountain masses.

BEAUTY, WONDER, WISE HUSBANDRY
National Parks, and Monuments and Forests

The United States has set aside a number of areas as National Parks and National Monuments to be preserved for the education and enjoyment of the people. The National Parks are remarkable chiefly for their scenic beauty or grandeur, and many are also sanctuaries for wild life. Some contain buffalo, mountain goats, elk, bears and other creatures which might otherwise become extinct. These parks, twenty-nine in number, total more than nineteen thousand square miles in area. The National Monuments, eighty-five in number, total over fourteen thousand square miles in area. In addition, five Historical Parks, eleven Military Parks, two Battlefield Parks, ten National Cemeteries, ten National Memorials and several miscellaneous areas are under the Department of the Interior. The National Forests, formerly called Forest Reserves, are administered by the Department of Agriculture, both to protect and develop timber resources and also to protect watersheds. The parks, monuments and forests receive every year thousands of visitors from all parts of the United States.

BEFORE the adoption of the Articles of Confederation several of the states claimed unoccupied lands west of the Alleghanies. These they ceded to the national government; and as different additions have been made to the United States, the ownership of all land (except in Texas) not already in private ownership has been vested in the general government.

Most of these lands have now passed into private ownership. However, the government still owns millions of acres scattered through many states of the West and some of it is open to settlement (though all the best agricultural land is gone), but much of it has been set aside for the use of all the people. These government reservations may be conveniently classed as National Parks, National Monuments and National Forests.

A National Park is created by Congress for the sake of developing and perpetuating it for the public enjoyment, and a National Monument is proclaimed by the President to conserve some restricted area of unusual scientific or historic interest. A National Forest, on the other hand, is administered by the Federal Government for the service of the public, in part to conserve the timber crop and in part to protect the watersheds and so prevent floods and resultant drought and aridity. There are 151 national forests with a total of 180,758,433 acres or 282,435 square miles. The summary at the end of the chapter indicates the size of each of the National Parks ranging from over 3,000 square miles to only a few acres. Most of the National Monuments are small, though some include large areas. It is probable that there will soon be a reclassification transferring some from one list to the other. The present arrangement has grown up without much plan, and some in each group logically belong in the other. Parks, monuments and forests welcome hundreds of thousands of vacationists every summer.

A map of the national forests would show vast areas reaching from Canada to Mexico along the rocky backbone of the continent and from the desert to the mountain meadows. Most of the national parks and nearly all of the national monuments would show up as mere dots here and there, although Yellowstone Park has an area of 3,458 square miles, chiefly in Wyoming, though spreading into Montana and Idaho; and Glacier Park on the Canadian border is nearly half as large. Several others are of considerable size. Exact figures are given at the end of the chapter.

The first of the national parks (barring the mile and a half of Arkansas Hot Springs) was that of Yellowstone Park, which was created in 1872. All the national parks are administered by the Department of the Interior under a National Park Service created in 1916.

BEAUTY AND WONDER AND WISE HUSBANDRY

The Nation's Pleasure-grounds

The national parks may be roughly classified as of four kinds—(1) those remarkable chiefly for their extraordinary scenic beauties, as stupendous waterfalls, gigantic trees of prehistoric age, the highest mountain peak in North America and the marine vistas of historic Acadie; (2) those displaying such evidences of erosion as remarkable limestone caves and stupendous canyons wonderfully carved and colored; (3) those illustrating glacial action; and (4) those containing volcanic phenomena, geysers and hot springs. The names, location and area of these parks are given in the summary, together with the dates of their creation. Instead of describing them in their chronological order, it seems more interesting to group them according to characteristics. The nation has provided roads, trails, supervised camp grounds and hotels.

Yosemite Valley, the best known feature of Yosemite National Park, cut by the Merced River and by glacial action, is a canyon seven miles long with walls in places three thousand feet in height down which pour the world's highest waterfalls. The Yosemite Falls is shown on page 394. Vernal Falls are unsurpassed for sheer loveliness. The Merced River here descends for 320 feet in a sheet of jade-green water to foam white among the bowlders at its foot. The trail from the valley rim leads one down among the spray-wet cliffs through a veritable rainbow which, every afternoon the sun shines, seems to follow one from step to watery step. Bridal Veil Falls, aptly named, drops 620 feet and the slender Ribbon Falls makes a straight drop of 1,612 feet. Nevada Falls drops 594 feet behind the evergreens. Equally spectacular are the summits that rise from the valley floor. Cathedral Rocks, El Capitan and Sentinel Dome are exceeded by Half Dome, which towers a sheer 4,892 feet, and Clouds Rest, 5,964 feet.

A Land Beloved of John Muir

But Yosemite Valley occupies only eight square miles out of a total of over eleven hundred which constitute the park. Above the valley's rim lies a region less well known save to groups like the Sierra Club and to individuals like John Muir, its first president, because well marked trails, canvas lodges and a motor road have only recently made it easily accessible. Lying on the western slope of the Sierra Nevada Mountains, Yosemite Park reaches Mount Lyell, the crest of the range, and the waters which feed the falls take their rise in the eternal snows. Glacier Point, on the valley rim, gives one a panorama of domes and pinnacles unsurpassed for its loveliness as the fingers of sunrise touch each in turn with gold.

The little known Waterwheel Falls of the Tuolumne River leap "high in the air in wheel-like whirls." The explanation of these falls is that the river, rushing down its canyon, encounters shelves of rock projecting from its bottom and throws enormous arcs of solid water upward, in some cases in a fifty-foot arc. One can but mention the government ranger-naturalist talks, the half-tame deer and brown bears, the carpets of wild flowers, the snow plant that pushes up like giant red asparagus, and the nightly bonfires that shower sparks to the moon. Yosemite Valley was discovered in 1851 by the Mariposa Battalion while pursuing Indians but was for long unknown save to miners and surveyors, soldiers and sheepherders.

Great Sequoias Saved from the Ax

Although Yosemite National Park includes a large grove of "big trees" (the Mariposa Grove) and Kings County National Park to the southward preserves a mammoth one in its extensive General Grant Grove Section, it is Sequoia Park which is most noted for conserving these oldest and biggest living things. The Sequoia gigantea, big cousin of the coast redwood or Sequoia sempervirens, is for the most part set apart in the Giant Forest of Sequoia Park. Before 1916 these trees were the property of individuals, but were purchased by Congress, aided by the National Geographic Society, and so saved from the possibility of falling before the

STEADY, THERE! It is a sheer drop of 4,892 feet from the summit of Half Dome where hikers in Yosemite National Park have an inspiring view of the valley of the Merced River.

lumberman. The oldest of these forest giants is unquestionably between three and four thousand years old, several hundred of them rise to three hundred feet in height and large numbers measure from twenty-five to thirty-seven feet at their base.

Brown bears, shy by day, roam the Giant Forest and sometimes raid campers' larders in the wee sma' hours. One moonlight night a two-yearling cub was seen running away on his hind legs hugging to his chest an outsize fruit cake, pan and all, for which he had overturned the kitchen cabinet of someone's motor camp; and several of the rascally fellows got into the Sierra Club commissary department and were routed in a smother of flour and a trail of bacon rinds. At that same camp a three-prong buck used to beg the campers for melon rinds. The wild life which is protected in all of the national parks is a unique source of entertainment.

A Land for Pack-horse Trips

From Giant Forest eastward up the forested slopes of the Sierras, Sequoia Park has been extended to include Mount Whitney, 14,496 feet, which drops almost sheer on its eastern front into the desert just north of Death Valley. To reach it, pack-horse campers negotiate the canyons of the Kaweah, Kern or Kings rivers, a wild region of castellated peaks, where brief afternoon thunder-showers brighten the aromatic conifers, and sun-baked middays are succeeded by chill nights. The Kern, unlike most Sierran streams, flows southward and its glacial-hewn canyon embraces more than forty peaks over 13,000 feet in height. The neighboring Kearsarge and Junction Passes were used by the California Forty-niners. Mount McKinley National Park in Alaska contains the one peak in the United States that towers higher than Mount Whitney —20,300-foot Mount McKinley, climax of an ice-coated range more fully described in the chapter on territorial possessions. Three scenic parks are in the East. Acadia, in Maine, was first established, then the Great Smoky Mountains in North Carolina and Tennessee, and the Shenandoah in Virginia.

Discovered by Champlain

Acadia (formerly Lafayette) National Park, occupies old French territory on the coast of Maine, with the ancient Mount Desert Mountains as its nucleus. "L'Île des monts deserts" (meaning, not "barren," but "wild and solitary") was discovered by Champlain in 1604 while exploring to the southward of De Monts' colony at the mouth of the Bay of Fundy. The Island, Mount Desert, was in 1688 presented by Louis XIV to the Sieur de la Mothe Cadillac, who left it for his governorship of Louisiana; and in 1713 the French king was obliged to cede this part of Acadia to England. After the capture of Quebec the island fell to the lot of the Province of Massachusetts, but Massachusetts gave it to Sir Francis Bernard; and although the property was confiscated during the Revolution, Bernard's son later secured a half interest in it—and sold it to American settlers. The other half, Marie de Cadillac, granddaughter of the original owner, regained for her family, but sold it bit by bit.

Acadian Woods and Waters

On this island, long inaccessible, fishing hamlets sprung up and the felling of the giant pines vied with the lobster industry. No steamer came until 1868. In the meantime, a few people of means discovered its delights of boating, climbing and buckboarding, and it became a favorite summer haven. Now the lands composing the park have been given to the nation from various sources. Though not yet fully developed, the Great Smoky Mountains Park, and the Shenandoah offer some of the most beautiful and impressive mountain scenery in the country, though the mountains do not tower so high as those in the West. The largest remaining hardwood forest is in the Great Smokies.

Of the parks remarkable as works of erosion (Zion and Bryce canyons, which are shown in pictures, Wind Cave and

ALONG THE BRIGHT ANGEL TRAIL IN THE GRAND CANYON OF ARIZONA

Travelers may reach the floor of the brilliantly colored Grand Canyon by a nine-mile ride on muleback down the steep Bright Angel Trail. It starts out from the south rim of the canyon.

TRANS WORLD AIRLINES

MONTEZUMA CASTLE, national monument in Arizona. The five-story, ash-pink adobe structure, 40 feet high, is one of the best preserved of the prehistoric cliff dwellings. It was dug into a natural cave at the top of a sheer cliff 145 feet high. White visitors who first saw the place incorrectly associated it with the Aztec Emperor of Mexico, Montezuma.

the cavelike shelters of the Mesa Verde cliff-dweller ruins), the Grand Canyon of the Colorado in Arizona is by far the most extraordinary. On viewing a sunset from Pima Point, a noted traveler and writer once said, "Peaks will shift and glow, walls darken, crags take fire, and gray-green mesas, dimly seen, take on the gleam of opalescent lakes." We depend upon the illustration on page 395 to give an idea of its weird carving and gorgeous coloring. Throughout the ages the Colorado River and its tributaries have gouged out of the sandstone a network of mysterious chasms and at one point the water flows red-silted nearly six thousand feet beneath the canyon's rim. The great natural barrier is more than two hundred miles in length, but in places one may descend on mule-back by trails that loop in zigzags. The total area of the park is over a thousand square miles.

A Perilous Journey

A Hopi Indian legend says that the first human beings ascended from the underworld by way of the Grand Canyon. While a number of white men had already seen and reported on its grandeur, the canyon was not explored until 1869 when Major John W. Powell with nine men in four small rowboats set out on a journey down the Colorado River. Four men gave up the hazardous trip through turbulent water before the party reached the lower end of the canyon.

Mesa Verde, Colorado, is a green tableland on which Richard and Alfred Wethrell, searching for lost cattle in 1888, came upon a hidden canyon and discovered—in a shelf under the overhanging edge of the opposite brim—a prehistoric cliff-dweller ruin that they called Cliff Palace. In a neighboring canyon they discovered Spruce Tree House, another of the best-preserved prehistoric ruins in America. A quarter century later an exploration conducted by Dr. J. W. Fewkes of the Department of the Interior unearthed Sun Temple on a mesa opposite Cliff Palace. The latter is the largest of many cliff-dwellings, each of which had living and storerooms for numerous clans, as well as kivas or rooms for religious ceremonials. The park was created in 1906.

Streams of Boiling Water Erupt

The national parks distinguished first for their volcanic origin include Hawaii, with two active volcanoes, a lake of boiling lava and an extinct volcano (described in the chapter on these islands), Crater Lake and Yellowstone Park. This park contains more geysers than are found in the rest of the world put together. Our five pictures of Yellowstone Park include one of the canyon through which the Yellowstone River foams. Not far distant, along the Lamar River, and elsewhere, there are fossil forests. Yellowstone is also one of the largest wild life refuges in the world. Some of the black bears actually permit automobile tourists to feed them and grizzlies come nightly to the garbage dumps. There are herds of elk, deer and antelope, moose and bison, buffalo, eagles and mountain goats. The buffalo were at one time in danger of complete extermination, as the settlers' fences cut off their pasturage and as the coming of the railroads caused them to be increasingly slaughtered. Now tourists riding quietly a little off the beaten trail often see a line of sentinel bulls rising black against a hill crest. Ranger-naturalists take parties on lecture-walks or talk about the nightly bonfires. The first white man who recorded a visit to the Yellowstone was John Colter, a member of Lewis and Clark's Expedition in 1807. Joseph Meek, W. A. Ferris, Father De Smet and James Bridger also told of it, and a government expedition was sent to explore it in 1859; but it was not until a large expedition went out in 1870 under H. D. Washburn and N. P. Langford that public incredulity was overcome and steps taken to create a park of the area.

One of the World's Deepest Lakes

This largest park was preceded by forty years by the smallest, the radioactive Hot Springs of Arkansas, which in De Soto's time the Indians constituted a Land of Peace. Lassen Volcanic National Park

A MERRY DOG-SLED RIDE past the vertical face of Half Dome, in Yosemite National Park. The park is open all year round and attracts numbers of winter-sport fans.

ELIZABETH HIBBS

YELLOWSTONE'S FAMOUS "OLD FAITHFUL" SPURTS UP ON TIME

With almost clocklike precision this geyser sends a column of steam and water into the air at regular intervals to the delight of tourists and camera fans. Yellowstone National Park, high in the Rocky Mountains where Wyoming, Montana and Idaho meet, has some three thousand springs, pools and performing geysers, as well as vast forest reserves for wild animal life.

THE GRAND CANYON of the Yellowstone and the Yellowstone River, as seen from Inspiration Point. The multi-colored cliffs are fringed with pines. Above the Upper Falls (109 feet), the river runs but little below the surrounding country, while after leaping the Lower Falls (308 feet), it rushes between canyon walls here more than a thousand feet high.

EMERALD SPRING is unlike the deeper springs of Yellowstone Park, which are a vivid blue. Practically the entire Yellowstone region is volcanic and contains numerous hot-water phenomena. Many hot springs bubble and throw water into the air several feet every minute or half minute and visitors to these must keep their distance, for in places the heat of the ground may be felt through the soles of our shoes, and the surface, yellow with sulphur, crumbles under our weight. Other phenomena of this strange Park include mud volcanoes.

in northern California was created just before the eruption of Lassen Peak in 1916. Once a row of fire mountains blazed along the mountains of the Pacific Coast states. Of these, Mount Mazama, in southern Oregon, underwent some terrific cataclysm in which the volcano fell into itself, jamming its vent and leaving a thousand-foot rim of cliffs about the cavity. In the ages that followed, cold springs poured in their waters until a lake covered all but the peak of one small cone. The result is Crater Lake, which varies from turquoise to blue-black; and the one tiny cone emerges as tree-clad Wizard Island.

The parks characterized by glaciers are Glacier Park and Rainier, while Rocky Mountain Park, in northern Colorado, and Grand Teton, in Wyoming, show signs of glacial action. Rainier rears its solitary white crown in Washington, where it towers 14,408 feet above Puget Sound, bearing a great cap of ice with ragged border. Its glacier system exceeds all others in the United States in both size and grandeur. Twenty-eight imperceptibly moving rivers of ice which have been explored and named, in addition to unnamed smaller glaciers, flow down its sides until their terminal moraines lose themselves in alpine fields of wild flowers. Rich forests of fir and cedar clothe the lower slopes, but from every open space on road or trail the great white dome glistens until one understands why the Indian name for deity, Tahoma (Tacoma) has been applied to it. Some prehistoric explosion has left a crater a mile wide in the mountain top, and the winds from the Pacific, suddenly cooled against its snow crown, deposit their moisture in terrific storms.

On Hoary-crowned Rainier

The snowfalls, settling in the crater, press themselves into ice and slide, of their own weight, down the rocky slopes, here grinding down the softer rock strata, there rumbling over precipices until the air of lower altitudes melts them to rivers milky with sediment. As there is less to impede the ice-flow in mid-stream, crevasses are formed which yawn, green and clear, for hundreds of feet, and climbing-parties are safe only with experienced guides.

Glacier Park has all of sixty glaciers, but is considered even more remarkable for the beauty of its rugged peaks and precipices and its several hundred glacial-fed lakes, the beds of which have been carved by glaciers of past ages. At Iceberg Lake, where there are miniature icebergs, even in midsummer, a glacier once hollowed a bowl beneath a rim of cliffs two thousand feet in height, and, curiously, another glacier hollowed a similar bowl so close on the other side of the mountain that, had they met, a mountain pass would have been created.

Amid Alpine Lakes

From the Continental Divide a dozen great valleys open gradually along the leisurely western slope, while seven drop abruptly on the east; and each of these valleys leads to some large lake. St Mary Lake and Lake McDonald, Lake McDermott with its minarets and Two Medicine Lake are too lovely for words, and even pictures leave out their aroma of spruce woods and the feeling of incredible freshness and soothing silence. Among the crags mountain sheep and goats watch while trout dimple the placid waters.

Rocky Mountain Park, "at the top of the world" in Colorado, was fathered by Enos A. Mills, and a huge glacier at the foot of a precipice of Long's Peak has been named for him. Many are the glacial-watered gorges. Those north of Long's Peak are called the Wild Garden and those south of the peak, Wild Basin. The many thickets of white-stemmed aspens make the region a favorite with beavers, who live on the bark.

Stalactites Like Stone Icicles

Caves as well as canyons have been formed by erosion, and numbers of limestone caverns have been preserved in the national monuments of the West. All have been formed by the action of under-

PHILIP GENDREAU

DEVILS TOWER, in Wyoming, is a great fluted column of volcanic rock. From the base it is 865 feet high. It tapers from a diameter of 1,000 feet at the bottom to 275 feet at the top.

COURTESY GREAT NORTHERN RAILWAY COMPANY

GLACIER NATIONAL PARK, in northern Montana, has more than sixty glaciers, and in its many basins, most of them carved by ice during the glacial period, lie clear blue lakes bordered by meadows of larkspur and Indian paintbrush. In places one can follow the old game trails of the Blackfoot Indians, some of whom still live in the vicinity.

CRATER LAKE lies in the Cascade Mountains of Oregon. The depression, which the lake occupies, is 2000 feet deep in places, and marks the place where the whole summit of the now extinct volcano (Mount Mazama) was engulfed by subsidence. Within its rim, the blue water of the lake mirrors clouds that shimmer in the setting of inverted evergreens, while Wizard Island and the islet called the Phantom Ship present pleasing varieties of color.

ground waters, perhaps through ages of time; and all are more or less characterized by stalactites formed by the drip of water impregnated with carbonic acid and by stalagmites formed by the ground splash from the stalactites which has caused them to grow slowly upward beneath them.

Visitors enter the Carlsbad Caverns in the Guadalupe Mountains of New Mexico through a musty-smelling hole that takes them directly from a noonday heat of more than a hundred degrees to an even, cool fifty-six. A trail almost a mile long down the main hall leads to a chamber called the King's Palace. Smooth paths, gentle stairways and electric lights, cleverly concealed behind rocks, make the going easy. Beyond the King's Palace, the Queen's Chamber and the smaller Papoose Room is the Big Room, a tremendous vaulted chamber more than half a mile long, six hundred feet wide and nearly three hundred feet high. There, in the Big Room, is Giant Dome, a leaning stalagmite thought to be more than fifty million years old.

Another wonder at Carlsbad is the swarm of millions of bats that billows from the entrance every summer evening at dusk. They consume tons and tons of insects before returning at dawn to their upside-down daytime life on the roof of a chamber hundreds of feet below the entrance.

Wind Cave of South Dakota and Mammoth Cave of Kentucky are two other national parks. The national monuments include several additional caves of great beauty—Jewel Cave in South Dakota, Lehman Caves in Nevada, the Oregon Caves, Shoshone Cavern in Wyoming and Timpanogos Cave of Utah.

The log cabin in which Abraham Lin-

THE GRANITE PEAKS of the Grand Tetons stand in scenic splendor beyond Jackson Lake in northwestern Wyoming. Sparkling glaciers lie in many of the ravines and gorges.

UNION PACIFIC RAILROAD

THE MILWAUKEE ROAD

MOUNT RAINIER explorers take time out for a breather before hiking up Paradise Glacier, part of the ice sheet that envelopes the peak rising 11,000 feet above the forested base.

coln is believed to have been born, now enclosed in a protective memorial building, is typical of national monuments to great Americans. In Washington, D. C., are memorials to Washington, Jefferson and Lincoln. These three presidents, along with Theodore Roosevelt, are also memorialized in the form of huge figures carved on the face of Mount Rushmore in South Dakota.

Many famous battlegrounds—Yorktown, Gettysburg, Chickamauga, Richmond and Little Bighorn, to name only a few—are marked by parks, cemeteries and monuments. One of these monuments is Appomattox Court House in Virginia where General Robert E. Lee's surrender to Ulysses S. Grant in 1865 brought an end to the bloody Civil War. Some other historical sites that have been set aside as national monuments include that of the first permanent English settlement at Jamestown, Virginia; Fort McHenry in Maryland, where the national anthem was inspired; Fort Laramie, Wyoming, which guarded the Oregon Trail; and Kill Devil Hill in North Carolina, scene of the world's first successful airplane flight, by Wilbur and Orville Wright.

When the United States Government first awoke to the need of forest conservation, an act of 1891 gave President Harrison authority to set aside ungranted land as forest reserves. The Yellowstone Park Timberland Reserve became the first of a series of such reserves. In 1905 the Forest Service was organized under the United States Department of Agriculture. In 1907 the name Forest Reserves was changed to National Forests. President Theodore Roosevelt showed special zeal in adding forests to the conservation areas.

The national forests of the United States and its territories cover approximately 180,000,000 acres—more than one acre for each inhabitant, on the average. Each year they yield millions of dollars to the Federal Government, chiefly from their timber crop, partly

UNION PACIFIC SYSTEM

YOSEMITE FALLS are among the highest in the world. If we add intermediate cascades and rapids the total descent is half a mile, and of this distance Yosemite Falls alone, in one sheer drop, accounts for 1430 feet. At a distance the roar of the water sounds like a train. There are a number of celebrated waterfalls in the Yosemite National Park in California.

COURTESY UNION PACIFIC SYSTEM

THE GRAND CANYON of the Colorado is here shown from the north rim. Here in the Kaibab National Forest are deer and mountain lions. The Canyon walls are higher on this side, resembling an intricate range of mountains on the border of a high plateau. Theodore Roosevelt once declared this view "absolutely unparalleled throughout the wide world."

CARLSBAD CAVERNS NATIONAL PARK

TEMPLE OF THE SUN, one of the magnificent limestone chambers in the Carlsbad Caverns of New Mexico. It is remarkable for its delicately colored stalactites and stalagmites.

A TREE GROWS IN MARIPOSA GROVE—AND GROWS AND GROWS!
Nowhere does man feel more insignificant than amid giant sequoia trees such as those above in Yosemite National Park, California. Fully grown trees average about 275 feet in height and 25 feet in diameter. Some have been known to grow between 350 and 400 feet tall and 96 feet in circumference. One stump, converted into a dance floor, can hold 40 or more people.

COURTESY UNION PACIFIC SYSTEM

ZION NATIONAL PARK, Utah, contains many fantastically eroded canyons. The most colorful is Zion Canyon, a fourteen-mile gorge in the Kolob Plateau, where a wedge-shaped opening only a few feet across, widens to a mile and deepens to three thousand feet. The Great White Throne is the name of the extraordinary formation shown above.

BRYCE CANYON National Park, Utah, resembles a giant amphitheatre, two miles by three, and a thousand feet in depth, the rim of which is eight thousand feet above the level of the sea. The softer parts of the plateau have been eaten away, leaving this semblance of towers and minarets, fortresses, pagodas, castles and cathedrals

BEAUTY AND WONDER AND WISE HUSBANDRY

from the leasing of grazing and other privileges. Yet fire annually destroys over a billion board feet of timber; insects, disease and occasional windfalls an even greater quantity. The fire-fighting led by the Forest Rangers, the saving of trees from loss and the planting of new growth is therefore of extreme importance; for the chief purpose of the forests is timber production. But the rangers also serve the public by advising as to camping, supplying information to lumbermen and what not. When the Forest Service took charge of forest lands, unregulated grazing was proving seriously injurious both to the growth of young timber and to the water supplies. Now, at the same time that the ranges are being brought back to their full productive power, the pasturage is being fully utilized according to the kind of range best suited to each kind and size of herd or flock. The Forest Supervisor is in charge of a property which must be protected, developed and improved; but he is also a sales manager and his responsibilities include fire protection, forest experiment stations and tree nurseries, forest products laboratories and the enforcement of grazing and lumbering privileges. Although the national forests are widely distributed, the states which include the largest areas of national forests within their boundaries are, in order, Idaho, California, Montana, Texas, Colorado and Arizona. The extensive west coast forests range in character from the giant Douglas firs of the humid northwest to the tall yellow pines of the semiarid southwest and include many species.

NATIONAL PARKS AND MONUMENTS: FACTS AND FIGURES

NATIONAL PARKS (28 in number)
(Location, date established and area in acres)

Acadia, Maine (1919), 29,978.08; Big Bend, Texas (1944), 692,304.70; Bryce Canyon, Utah (1928), 36,010.38; Carlsbad Caverns, N. M. (1930), 45,846.59; Crater Lake, Ore. (1902), 160,290.33; Everglades, Fla. (1947), 1,258,361; Glacier, Mont. (1910), 999,015.15; Grand Canyon, Ariz. (1919), 645,295.91; Grand Teton, Wyo. (1929), 299,580.45; Great Smoky Mountains, N. C.–Tenn. (1930), 507,159.16; Hawaii, T. H. (1916), 179,950.90; Hot Springs, Ark. (1921), 1,019.13; Isle Royale, Mich. (1940), 133,838.51; Kings Canyon, Calif. (1940), 453,064.82; Lassen Volcanic, Calif. (1916), 103,809.28; Mammoth Cave, Ky. (1936), 50,695.73; Mesa Verde, Colo. (1906), 51,017.87; Mount McKinley, Alaska (1917), 1,939,319.04; Mount Rainier, Wash. (1899), 241,571.09; Olympic, Wash. (1938), 887,986.91; Platt, Okla. (1906), 911.97; Rocky Mountain, Colo. (1915), 254,735.70; Sequoia, Calif. (1890), 385,178.32; Shenandoah, Va. (1935), 193,472.98; Wind Cave, S. Dak. (1903), 27,885.67; Yellowstone, Wyo.-Mont.-Idaho (1872), 2,213,206.55; Yosemite, Calif. (1890), 757,617.36; Zion, Utah (1919), 94,241.06. Total area, 12,640,364.64 acres.

NATIONAL MONUMENTS (85 in number; total area, 8,999,693.62 acres; location and date estab.)

Ackia Battleground, Miss., 1938; Andrew Jackson, Tenn., 1942; Appomattox Court House National Historical Monument, Va., 1940; Arches, Utah, 1929; Aztec Ruins, N. M., 1923; Badlands, S. Dak., 1939; Bandelier, N. M., 1916; Big Hole Battlefield, Mont., 1910; Black Canyon of the Gunnison, Colo., 1933; Cabrillo, Calif., 1913; Canyon de Chelly, Ariz., 1931; Capitol Reef, Utah, 1937; Capulin Mountain, N. M., 1916; Casa Grande, Ariz., 1918; Castillo de San Marcos, Fla., 1924; Castle Clinton, N. Y., 1950; Castle Pinckney, S. C., 1924; Cedar Breaks, Utah, 1933; Chaco Canyon, N. M., 1907; Channel Islands, Calif., 1938; Chiricahua, Ariz., 1924; Colorado, Colo., 1911; Craters of the Moon, Idaho, 1924; Custer Battlefield, Mont., 1946; Death Valley, Calif.-Nev., 1933; Devils Postpile, Calif., 1911; Devils Tower, Wyo., 1906; Dinosaur, Utah-Colo., 1915; Effigy Mounds, Iowa, 1949; El Morro, N. M., 1906; Fort Frederica, Ga., 1945; Fort Jefferson, Fla., 1935; Fort Laramie, Wyo., 1938; Fort Matanzas, Fla., 1924; Fort McHenry National Monument and Historic Shrine, Md., 1939; Fort Pulaski, Ga., 1924; Fort Sumter, S. C., 1948; Fossil Cycad, S. Dak., 1922; George Washington Birthplace, Va., 1930; George Washington Carver, Mo., 1951; Gila Cliff Dwellings, N. M., 1907; Glacier Bay, Alaska, 1925; Gran Quivira, N. M., 1909; Grand Canyon, Ariz., 1932; Great Sand Dunes, Colo., 1932; Homestead National Monument of America, Nebr., 1939; Hovenweep, Utah-Colo., 1923; Jewel Cave, S. Dak., 1908; Joshua Tree, Calif., 1936; Katmai, Alaska, 1918; Lava Beds, Calif., 1918; Lehman Caves, Nev., 1922; Meriwether Lewis, Tenn., 1925; Montezuma Castle, Ariz., 1906; Mound City Group, Ohio, 1923; Muir Woods, Calif., 1908; Natural Bridges, Utah, 1908; Navajo, Ariz., 1909; Ocmulgee, Ga., 1936; Old Kasaan, Alaska, 1916; Oregon Caves, Ore., 1909; Organ Pipe Cactus, Ariz., 1937; Perry's Victory and International Peace Memorial, Ohio, 1936; Petrified Forest, Ariz., 1906; Pinnacles, Calif., 1908; Pipe Spring, Ariz., 1923; Pipestone, Minn., 1937; Rainbow Bridge, Utah, 1910; Saguaro, Ariz., 1933; Scotts Bluff, Nebr., 1919; Shoshone Cavern, Wyo., 1909; Sitka, Alaska, 1910; Statue of Liberty, N. Y., 1924; Sunset Crater, Ariz., 1930; Timpanogos Cave, Utah, 1922; Tonto, Ariz., 1907; Tumacacori, Ariz., 1908; Tuzigoot, Ariz., 1939; Verendrye, N. Dak., 1917; Walnut Canyon, Ariz., 1915; White Sands, N. M., 1933; Whitman, Wash., 1940; Wupatki, Ariz., 1924; Yucca House, Colo., 1919; Zion, Utah, 1937.

Unspoiled Wilderness Lands

National Parks and Reserves of Canada

The Canadian Government, through the National Parks Service of the Department of Resources and Development, administers thirty-one parks and animal reserves set aside for the purpose of preserving the scenery, wildlife and historic sites of Canada. In these areas one may see the bears and mountain sheep of the Rockies, deer, moose, such fur-bearers as foxes, marten, beavers and others, in addition to a great variety of birds; while the lakes and streams abound with trout, pike, bass and other game fish. Fairly recently vast areas of heretofore uninhabited country have been made accessible by rail, motor roads and small water craft. The scenically beautiful regions thus set aside often lend themselves to winter sports as well as to summer camping and mountaineering. It is a region of unspoiled wilderness visited by vacation-seekers from both Europe and the United States as well as Canada.

THE Dominion of Canada has set aside more than 30,000 square miles as national parks for the use and enjoyment of the people. These areas, of varying sizes, have been established to maintain the scenic beauty and conserve the wild life of the regions in which they are situated, and in the case of eleven parks, to preserve historic sites of outstanding national interest. In Alberta are the Banff, Jasper and Waterton Lakes parks on the eastern slope of the Rocky Mountains; Kootenay and Yoho parks are on the western slope of the Rockies in British Columbia. Farther west in the Selkirk mountains of British Columbia are the Glacier and Mount Revelstoke parks. In Alberta are also four wild animal parks; Buffalo, near Wainwright; Elk Island, near Lamont; Nemiskam, near Foremost, and Wood Buffalo in the north. In Saskatchewan is Prince Albert National Park, and in Manitoba, Riding Mountain Park. While the greater number of Canada's national parks are situated in the western provinces, in recent years notable areas have also been set aside in Eastern Canada. Ontario possesses three national parks—Point Pelee, Georgian Bay Islands, and St. Lawrence Islands. In Nova Scotia are Cape Breton Highlands and Fort Anne parks; in New Brunswick, Fort Beauséjour and Fundy parks; and a park has been established in Prince Edward Island, on the Gulf of St. Lawrence.

To return to the interesting group of Rocky Mountain parks, the old fur-traders—dauntless explorers of the fur companies, seeking new lands from which to get furs or seekers of new territory to add to the British Empire—fought their way for a half a century across the passes of the Rockies to the Columbia. Alexander Mackenzie was the first of these who made his way to the Pacific, blazing the way for a line of trading-posts. Another outstanding figure was Simon Fraser. In 1814 Grabriel Franchère succeeded in crossing Athabaska Pass discovered in 1811 by David Thompson, and from that date forward the fur brigades of the British companies went jingling twice a year along the Athabaska Trail. Yellowhead Pass, a point on the Great Divide at which the mountains may be crossed at 3,700 feet, became by 1826 a route to the Columbia via the Fraser River. Gold-seekers came to British Columbia after 1860. Finally came the railroad surveyors; and to-day the motorist can drive over the Banff-Jasper Highway which was opened for travel in 1940.

The four parks that lie together along the Rockies—Jasper, Banff, Yoho and Kootenay—form an area three-fifths as large as Switzerland; and of these, the largest is Jasper. There are two main approaches to Jasper; on the east from Edmonton up the Athabaska Valley, on the west by Yellowhead Pass and the

Miette River. Hundreds of the peaks within park boundaries have neither been named nor climbed. Indeed, the rugged northern portion of Jasper Park is still unexplored save by aeroplanes, which have seen long, shining glaciers and waterfalls that leap into steep black canyons. The ranges lie parallel, wave on wave to the westward, like giant combers; but on the east the mountains rise abruptly from the plains to altitudes of six or seven thousand feet, with long valleys running between them. On the east slope of the Rockies there is little rainfall and the air is electric, but the western slopes are well watered.

GOVERNMENT OF ALBERTA

TWO YOUNG EXPLORERS in Jasper Park trade stares with a herd of wild mountain sheep.

Although Jasper has a hundred and fifty miles of motor roads, increasing numbers of tourists in hob-nailed boots climb the trails, perhaps with camp outfits and Swiss Alpine guides, or ride the sure-footed mountain ponies through spruce woods and across flowery meadows to the very foot of the snow fields, putting up at camps established near several points of interest, if they cannot make their hotel by nightfall. One may, in the course of a camping-trip, visit the Columbia Ice Field, where there is a blanket of about 110 square miles of ice to mark the divide from which streams flow in three directions—to Hudson Bay, the Pacific and the Arctic oceans. Or one may make a trip to Mount Robson in the Provincial Park of that name which adjoins Jasper, rising to 12,972 feet above a sea of peaks on the park boundary. From Kinney Lake the trail climbs to the Valley of a Thousand Falls, which leap noisily from the melting ice fields of late summer. At Berg Lake, high on the mountainside, we see Berg Glacier hanging blue and clear above the lake and dropping chunks of ice.

Before leaving this park one should follow the foaming Maligne River Canyon and see glacier-fed Maligne Lake, the largest in Jasper, as well as topaz Chrome Lake, opal-tinted Edith Lake and blue Pyramid and Patricia lakes. One should have a glimpse at least of some of the canyons, Ogre, Athabaska, Fiddle Creek or those less well known. Then a last look about as we leave Jasper townsite, to the east at Old Man Mountain (Roche Bonhomme), lying like a sleeping warrior along the Colin Range; north to the reddish rocks of Mount Pyramid, west to the pine-clad Whistlers and pyramidal Mount Fitzwilliam; then to the lovely snow-crowned peak that dominates all the others, Mount Edith Cavell. If the cloud wreaths that often encircle it permit, one may see on its bosom a mammoth glacier said to resemble the outspread wings of an angel. On its left stands Signal Mountain, from which the fur-traders once watched for the approach of pack-trains from the Pacific.

Yoho National Park was opened to tourist travel in 1927 through the completion of the highway known as Kicking Horse Trail. When Sir James Hector, geologist of the Palliser Expedition, after having discovered what we know as Banff, discovered the pass later used by

CANADIAN NATIONAL RAILWAYS

A LONE FISHERMAN tests his skill and luck in Jasper Park. Snow deposits on the weathered battlements of the Ramparts are reflected in the clear water of Amethyst Lake.

the first transcontinental railway of Canada, his Indians named the river that the trail follows in commemoration of the episode of Dr. Hector's saddle-horse kicking him in the chest. The difficulties of those days are suggested by the fact that one of the pack-horses fell over a 150-foot slope, landed on his back in a tree and finally was brought up astride a great trunk, comically helpless. The building of the motor road was likewise precarious, for logs had to be lowered 1,200 feet by drum and cable; and had a bowlder been dislodged, it might have wrecked a train on the tracks below. The highway resembles a crease gouged in the side of a cliff, and forms one link in a loop of some 275 miles with the Banff-Windermere Highway and the Columbia Highway. This is one of the most spectacular motor roads to be seen anywhere. The railroad above mentioned was completed in 1885, and trains must rise 1,500 feet in the sixteen miles from Field to Lake Louise.

Among the beauty spots of Yoho Park are Emerald Lake, Lake O'Hara, Wapta Lake and the beautiful Yoho Valley, down the sides of which cascade dozens of beautiful waterfalls, from the great ice fields above.

This Yoho Valley is one of the beauty-spots of the Rockies. It is about fourteen miles long and a mile deep. Perhaps the most noted of all the waterfalls is Takkakaw which has its source 2,500 feet above the floor of the valley, and finally tumbles in a cloud of milky green water into the Yoho River. The word "Yoho" is an Indian exclamation meaning wonderful.

Kootenay National Park was formed by setting aside a strip five miles wide along each side of the road from Vermilion Pass, on the intramontane highway through Banff, to Sinclair Canyon. Thus it preserves the scenery along this first motor road to be constructed through the central Canadian Rockies. It crosses Vermilion Pass, which likewise was discovered by Dr. Hector. It was named for the red oxide (vermilion) made from the red ochre of the region by the Kootenay Indians. The park has Radium Hot Springs and a government swimming-pool.

Seasonal motor licenses entitle the tourist to thirty days free camping at any of the recognized camp-sites within Jasper, Yoho, Kootenay or Banff National

Courtesy Commissioner of National Parks
MOUNT EDITH CAVELL

The serene beauty of this mountain, named for the heroic nurse, is here seen superbly mirrored in a rarely lovely green lake of the same name. It is a centre of interest in Jasper Park.

parks. These camp-sites are equipped with shelters, stoves, tables and other conveniences—likewise with signs significantly warning the camper not to tease the wild animals! Provision has also been made for motor trailers.

Banff and Lake Louise are world-famous for their scenic beauty. The first reservation around the hot springs at Banff was made in 1885 with the completion of the first railway across the Rockies. Seventeen years later Banff

Park was enlarged to five thousand square miles, but has since been reduced to its present proportions of 2,564 square miles. A party of those early railroad surveyors first investigated when they saw steam issuing from a hole in the mountainside. Setting up a pole with cross-pieces to make a ladder, they discovered a forty-foot cave arched over a hot pool fed by subterranean springs. Four other radio-active mineral springs were later discovered, and the total flow of the five has been estimated at a million gallons a day. At one of these springs the government has erected a fine public bathhouse.

Every tourist visits perfect little Lake Louise, with its blue depths set against the background of Victoria Glacier. It can be reached by motor road from Lake Louise station. Banff, nestled in an amphitheatre of mountains but itself 4,538 feet high, in the green valley of the Bow River, is cool and dry and balsamic. There those inclined may put up at good hotels or cottages or at the government camp-site, and golf among the clouds, inspect the zoo and the museum, attend the Indian Day celebration, at which the aborigines parade and dance in costume, or come for winter sports, with the trees dripping icy stalactites and silvered with frost, and the great peaks glittering. The skiing, snow-shoeing, skating, ice-boating and tobogganing come to a climax, in a winter carnival. Nearly two hundred miles of motor roads centre around Banff, while the park contains some seven hundred miles of trails. One should visit Lake Minnewanka (Spirit Water) and fish from a boat at 4,800 feet elevation, and see Castle Mountain, with its high turrets and natural drawbridge, which legend claims as the home of the Chinook Wind—harbinger of spring. Then there are lakes too numerous to mention, the Mistaya, Hector, Twin and Bow, the latter 6,500 feet toward the clouds. And there are wonderful mountains—four great groups of them in the southern portion of the

THE JASPER-EDSON HIGHWAY, in Jasper National Park, Alberta, parallels the course of the Athabasca River—once an important trade route of the Hudson's Bay Company.

PHOTOS, CANADIAN GOVERNMENT TRAVEL BUREAU

THE TOWN OF JASPER in Jasper National Park. Though the town itself is high, it almost seems to be lowland in contrast with the surrounding peaks, which loom much higher still.

ABOVE SAPPHIRE LAKE O'HARA, Yoho National Park, looms Mount Cathedral, the pinnacle at right.

CANADIAN PACIFIC RAILWAY

a snow field of the Selkirks, it measures 3,600 feet. This great glacier is melting faster than it grows, and in course of time will probably disappear. However, as the average rate of retreat, as marked by the red point on the mountain walls, is only about thirty-three feet a year, it will be a long time before it is visibly smaller.

From Glacier there is a trail to Cougar Valley and Nakimu Caves. These caves contain one perpendicular rise of eighty feet, up which a flight of steps has been built. In Cougar Valley the stream suddenly disappears, to emerge about 450 feet farther on, and to disappear twice more in the course of a mile. From its underground channel weird rumbling noises sound in the caves, of which there are over a mile, some of them incrusted with pale pink limestone cauliflowers. Glacier Park is a favorite with Alpine climbers.

Mount Revelstoke National Park, seemingly at the top of the world, is the scene of an annual ski-jumping contest at which the world's record has been closely approached. Jumps of 269 and 287 feet have been made in recent years. There is a fourteen-mile drive from the mountain hamlet of Revelstoke in which the road, formally opened in 1927 by the Prince of Wales, zigzags through virgin forest or along a rock ledge up the face of Mount Revelstoke, ascending 4,515 feet. This drive gives one a panoramic view of the valleys of the upper Columbia and the Illecillewaet, the Clach-na-Cudainn Icefield, Lake Eva and Lake Millar, and of Summit Lake gleaming at the top of Eagle Pass. In the high meadows, one may catch glimpses of caribou and wade to one's knees in wild flowers. Around sundown the snow crowns of the Selkirks and

park alone. Climax of all, there is the towering sharp peak of Mount Assiniboine 11,860 feet high and perhaps 1,500 feet above any of its neighbors. The first white men to achieve the summit were Sir James Outram and two Swiss Alpine guides. One at a time, on a rope held taut by the other two, they crawled to the actual tip of the mountain, and peered down a sheer wall to a glacier lying six thousand feet below.

Glacier, another of the mountain parks, is not reached by automobile, but by rail, or by pack-train. Here the Illecillewaet (Indian for "Swift Water") hurries through a green valley. A glacier of the same name makes a spectacular icefall against dark fir woods. Falling from

CANADIAN PACIFIC RAILWAY

MOUNTAIN PARADISE near Hector village in Yoho park. Between Hector and Field are the famous Spiral Tunnels, where rail lines make almost complete circles inside the mountains.

NATIONAL PARKS BUREAU, OTTAWA

THE PYRAMID-LIKE APEX OF LOFTY MOUNT ASSINIBOINE

The sheer summit rises in aloof grandeur above treacherous ice fields, an eternal challenge to the skill and endurance of mountain climbers. Mount Assiniboine, 11,870 feet high, is in Banff National Park and was first scaled in 1901, by Sir James Outram. The forbidding peak, though not the highest in the area, is often called the "Matterhorn of the Rockies."

Courtesy Commissioner of National Parks

THE GIANT'S STEPS IN PARADISE VALLEY

Paradise Valley is a lovely stretch of open country, about six miles long, between the Valley of the Ten Peaks and Lake Louise in Banff National Park. From Horseshoe Glacier, sweeping across at its head, clear streams flow down and come together to form Wastach Brook, which, midway of the valley, drops in a series of graceful cascades over a giant's stairway of rock.

UNSPOILED WILDERNESS LANDS

Courtesy Commissioner of National Parks
A ROCKY MOUNTAIN RAM
Even about the motor highways of Banff Park the Bighorn or Rocky Mountain sheep can be viewed at close range, so tame has this keen-eyed "chamois of the American West" become.

holds almost twenty feet of snow.

Waterton Lakes is the last national park in the high mountains. Scenically, it is one with the United States Glacier Park just over the International Boundary, and in 1932, with Glacier Park was proclaimed the Waterton-Glacier International Peace Park. Indeed, the big Waterton Lake lies in both. Of all the splendid reservations set aside by Canada, none is lovelier than this combination of gleaming snow peaks, with their rivers of ice and their vivid blue lakes set four thousand feet high beneath slopes of gray limestone curiously banded with red. The park begins at the crest of the Great Divide on the west and descends the wooded slopes of the Rockies to the rolling foothills of Alberta. The lake was named for an English naturalist and pioneer in wild-life conservation. The first white settler, John George Brown, an Oxford man, came here in 1865 and eventually became acting park superintendent. The neighboring Blackfoot Indians are often seen about the park. The tourist will find fully 150 miles of trails as well as launch routes which make it easy to

their attendant cloud-wraiths light up with a rose alpen-glow.

Near the summit of Mount Revelstoke is a rocky cleft a hundred feet long and twenty wide, known as the Ice Box. In the hottest weather of mid-summer it

Photograph by Byron Harmon
CASTLE MOUNTAIN AS SEEN FROM VERMILION SUMMIT
Halfway between Banff and Lake Louise in Banff National Park is this titanic fortress, with a foundation over eleven miles long and walls towering a mile above the base. To the east of it, the strata of limestone forming the mountains are almost turned on end; westward, they lie horizontally, giving the summits a more domelike shape.

Courtesy Department of Immigration and Colonization

LAKE LOUISE, WHERE BEAUTY SMILES ETERNALLY

Ages ago a glacier hollowed a deep basin where now in a "hanging" valley rests one of the most exquisite little lakes in all the world. Over its cold waters play in infinite variety the hues that shimmer on the dragon-fly's wing and in the peacock's tail, with that nameless blue that arises only from a glacier's heart.

Canadian National Parks Bureau
RIDING A TRAIL AT UPPER WATERTON LAKE
One of the favorite trails of visitors at Waterton Lakes National Park in Alberta leads along the beautiful Upper Waterton Lake, which is shown above, with a view of Mount Cleveland's snow-clad summit in the distance. Upper Waterton Lake lies across the International Boundary line separating the United States from Canada.

explore lake and forest in all directions.

The Park headquarters are at the new town of Waterton Park where are also hotels, boarding houses, stores and other conveniences for tourists. There are excellent tennis courts, a golf course, and the bathing is good. A large steamer makes daily excursions across the boundary line into Glacier Park.

There are many miles of fine motor roads in the Park, but one of the most satisfactory methods of seeing the scenic wonders is by saddle pony over the mountain trails. Some of these trails lead to the top of some of the lesser peaks. Mountain climbers enjoy the ascent to many of the peaks.

Wild life in the Park is abundant. Rocky Mountain sheep and goats, elk and mule-deer and bears are numerous. Many species of fish, particularly trout, are found in the lakes, and fish hatcheries are constantly adding to the numbers.

The new Prince Albert Park, in northern Saskatchewan, with its thousands of shining lakes and connecting streams, is a paradise for canoeists during its brief summer warmth. One can make a circular canoe trip of close to a hundred miles, with a few short portages, where canoe and duffle must be carried along a trail often but two hundred yards in length and soft with pine needles beneath one's moccasins. One paddles through tunnels of fragrant jackpines, white birches and white spruces which grow down to the margins of little beaches of white sand. In the chill waters, gray trout leap, moose swim across the lakes at twilight and deer step daintily into their margins to drink. Game birds nest beneath the saplings, pelicans and cor-

A GESTURE TO THE UNIVERSE FROM THE SIDE OF MOUNT ABBOTT

In Glacier National Park, British Columbia, are many reminders of the Alps, in its forests, its wildflower gardens and the lure of its rocky mountain tops. For those who seek adventure or who wish to study glacier formations it holds strong attractions. Near the top of Mount Abbott (8,081 feet) is an outlook which gives a fine view of Illecillewaet Glacier.

CANADIAN GOVERNMENT TRAVEL BUREAU

A SPINE-TINGLING RIDE on a chair lift carries a skier to the top of the run on Mount Norquay, near Banff. Range on range of the jagged Canadian Rockies are spread out below.

WATERTON TOWNSITE is on a spur jutting out into a lake in Waterton Lakes National Park, on the Alberta-Montana border. On the United States side is Glacier National Park.

PHOTOS, NATIONAL FILM BOARD

GOLFERS on the eighteenth hole of the course in Prince Albert National Park, Saskatchewan, line up their putts. The hilly terrain of the park makes for an ideal golf course.

BRITISH COLUMBIA TRAVEL BUREAU

SCENIC HIGHWAY running through a defile of overhanging rock in Kootenay National Park.

morants fish, and bears, looking like black stumps, forage in the berry patches.

Prince Albert Park, formally opened in 1928, is less than six hundred miles drive from Winnipeg. On Waskesiu Lake in the northern portion of the park, a crystal stretch of twenty miles, a fine campground site has been laid out. Kingsmere Lake is among the more significant of these northern waters. This region was once the hunting ground of the Cree Indians, and to this day they leave votive offerings—a pipeful of tobacco, perhaps, or an eagle's feather—on "Old Man Rock" in Waskesiu River, believed to be the special habitation of Wee-sa-ka-chack, a divinity possessed of the power to assume any form he chooses.

Manitoba's Riding Mountain

Riding Mountain National Park in Manitoba, which contains an area of approximately 1,148 square miles, is one of the newer additions to Canada's National Park system. The park is situated on a broad plateau which forms the summit of the Riding Mountain, nearly one thousand feet above the level of the great agricultural plains to the east, and forms a sanctuary for wild life, as well as an unrivalled summer playground. The park contains one of the largest herds of wild elk in Canada, and a herd of buffalo has also been placed in a large fenced enclosure near Lake Audy. Moose, deer, beaver and other kinds of mammals thrive in this natural environment, and many species of bird life, including waterfowl, are seen in abundance.

During the Ice Age, huge boulders were carried high up on the sides of Riding Mountain by the advancing glacial sheet. The boulders are there today, adding to the park's picturesqueness.

The park headquarters are situated at Wasagaming, a summer resort on Clear Lake, where many facilities have been provided for outdoor life and recreation. A fine bathing beach stretches for nearly a mile and a half in front of the townsite, and a large motor camp ground, equipped with rustic shelters, camp stoves, tables and firewood, is available to campers on pay-

GREEN GABLES farmhouse was made famous in a novel for young people by E. M. Montgomery. Anne's beautiful home is now the golf clubhouse in Prince Edward Island National Park.

PHOTOS, CANADIAN GOVERNMENT TRAVEL BUREAU

TRAVELERS in Prince Edward Island National Park stop to stretch their legs in the coastal village of North Rustico. Fishermen's nets are spread out on a rack to dry.

BUFFALO PRESERVE. One of the last great buffalo herds in North America roams in Elk Island Park, Alberta. National parks generally serve as sanctuaries for wildlife.

PHOTOS, ALBERTA GOVERNMENT

RECREATION, TOO! Bathing is another advantage offered by Elk Island Park. Sandy Beach is a popular weekend rendezvous for families in the vicinity who enjoy outdoor life.

ANTELOPE AT NEMISKAM   A DEER IN VELVET

Courtesy Commissioner of National Parks

ONE OF THE MOOSE IN BUFFALO NATIONAL PARK

Although the antelope have their special reserve at Nemiskam in southern Alberta, they are found in other regions as well. The deer shown above, one of the many that range over the parks, has a crown of new antlers which have not yet lost their covering of skin, the velvet. Moose, although lovers of the swamps, are increasing in several mountain park areas.

Canadian National Parks Bureau

A SUMMER DAY ON WASKESIU LAKE

Lake Waskesiu in Prince Albert National Park in Saskatchewan is a beautiful sheet of water, with long, safe, sandy beaches which form a delightful summer playground. The canoeist also finds the water of the lake almost perfect. Canada is fortunate in the number of attractive lakes which are to be found in almost every section of the Dominion.

ment of a nominal fee. A community building, museum and lecture hall, and bathhouses are open to the public without charge. Numerous trails and drives are also available for hiking, riding and motoring. The park is accessible by fine roads which link up with the provincial highway system.

The first of the Canadian wild animal reserves was Elk Island Park, to which one may drive from Edmonton. Originally (in 1906) designed for elk, and now containing moose and deer as well, it later added a buffalo herd. It has also become a sanctuary for game birds and waterfowl. This enclosure of fifty-one square miles contains about 2,500 buffalo; but, in 1908, Buffalo National Park was set aside near Wainwright and to this new area most of the first buffalo herd were transferred. Since then their numbers have increased considerably.

These great beasts, originally brought from Banff and Montana, were survivors of the tens of thousands that had roamed the plains of North America for many centuries and had been in serious danger of complete extermination. Their numbers have increased many thousands, and about 5,000 are kept in the park. Here in the rolling prairies of Alberta, sleek brown herds may be seen from the windows of the transcontinental trains as they graze on the sere long grasses, the little yellow calves hiding behind their mothers; and in autumn the shaggy-maned bulls paw up clouds of dust as they fight for leadership, bellowing resoundingly. Buffalo Park contains, besides these beasts, numbers of yaks, elk, mule-deer, moose, cattle and cattalo—a cross between buffalo and cattle. A high

THE WISHING WELL IN RIDING MOUNTAIN NATIONAL PARK

The rustic bridge crosses Bogey Creek and below it, to the right, is the wishing well, where wishes are supposed to come true. The lovely spot is on the north shore of Clear Lake, the largest—nine miles long and two miles wide—of a number of lakes in the park. Wild life is protected in the area, and one may see a wide variety of North American mammals.

CANADIAN GOVERNMENT TRAVEL BUREAU

LIMESTONE CAVES and formations are commonly found in the Georgian Bay Islands National Park, Ontario. The limestone tower from which plants are growing is called the Flowerpot.

wire fence and, outside it, a plowed fireguard protect the wild denizens of both Buffalo and Elk Island parks. Since their numbers have become so large, reductions are made by slaughter, and the meat and robes are sold.

The remaining two wild animal parks are Nemiskam in Alberta, near Foremost, and Wood Buffalo, a huge unfenced area in Alberta and Northwest Territories. In Nemiskam are pronghorn antelope, the delicate and lithe-limbed creatures that once inhabited the western plains of Canada and the United States in hundreds of thousands.

Three beautiful areas have been established as national parks in Ontario. These include Point Pelee National Park which stretches in the form of a large inverted triangle into Lake Erie. The park contains semi-tropical vegetation, has fine groves of trees and forms an ideal sanctuary for migratory waterfowl, which find rest and shelter on the large marshes in the central portion of the park. Camp sites have been laid out for visitors, and opportunities for bathing on the thirteen miles of sand beaches that lie off the eastern and western sides of the park are unexcelled.

The Georgian Bay Islands Park is formed by thirty islands in the Georgian Bay archipelago. The largest island, Beausoleil, has been developed as a camping resort and recreational area, with numerous docks and camp shelters available at points along the shoreline. Flowerpot Island, so named because of two stone pillars resembling enormous flower-pots, is an especially curious formation. Bathing, fishing and boating are among the sports to be enjoyed in the park.

The St. Lawrence Islands Park comprises a small area on the mainland and thirteen islands in the Thousand Islands of the St. Lawrence River, many of which have been developed as camping and picnic areas. As early as 1904, some of these islands were reserved for park purposes though the park itself was not established until later.

The most recently established units in Canada's national-park system are situated in the Atlantic Provinces. The Cape Breton Highlands Park, situated in the northern part of Cape Breton Island, Nova Scotia, owes its name to the well-wooded hills that rise sharply from the sea to a height of 1,200 to 1,700 feet, and resemble in appearance the Highlands of Scotland. Picturesque headlands jut out into the Gulf of St. Lawrence and the Atlantic Ocean, forming delightful bays and sandy coves, which are visible from a motor highway called the Cabot Trail, which follows the coastline for many miles. The country surrounding the park is very popular with tourists. The picturesque fishing ports and villages are centers of attraction for artists, and also provide fine opportunities for deep-sea fishing and boating. The park headquarters are located in the village of North Ingonish, situated north of the city of Sydney.

The Prince Edward Island Park includes a coastline strip along the northern

NATIONAL FILM BOARD

AN ATTRACTIVE GATEWAY, made of timber and stone, bids you enter Riding Mountain National Park. After a brief halt, you roll on through deep woods beside shimmering lakes.

AMONG THE BEAUTIFUL THOUSAND ISLANDS

Canadian National Parks Bureau

Between Kingston and Brockville the St. Lawrence is dotted with islands, a map as early as 1727. Some groups are named for soldiers, others for of varying size and shape. The name "Thousand Islands" appears on naval officers, others for Indians, and others for civilians.

shore facing the Gulf of St. Lawrence, embracing some of the finest beaches in eastern Canada. The surf-bathing is magnificent, and boating, fishing and hiking may also be enjoyed under ideal conditions. The park also contains the building known as Green Gables, made famous in the novel "Anne of Green Gables" by L. M. Montgomery.

The national historic parks of Canada include Fort Anne in Nova Scotia and Fort Beausejour in New Brunswick. Fort Anne National Park, containing an area of 31 acres, is situated at Annapolis Royal, the site of the first European settlement in North America, north of Florida, made by Champlain and deMonts in 1605. The old earthworks of the fort are in a fine state of preservation, and a large building erected in 1797 and recently restored, is used as a museum to house many interesting exhibits. It was erected for the officers of the fort by the Duke of Kent, father of Queen Victoria, while he was commander of the forces in North America.

Fort Beausejour Park situated near Sackville, New Brunswick, contains 59 acres surrounding the remains of a French fort erected prior to 1755, which was captured that year by the English under Monckton. In the park is a new historical museum, which contains many objects connected with the history of the Isthmus of Chignecto.

An important function of the National Parks Bureau is the preservation and marking of sites of national historic im-

Courtesy Canadian National Railways

A STEAMER IN THE RAPIDS OF THE ST. LAWRENCE

Among the Thousand Islands, the St. Lawrence is swift, and in many places dangerous unless the pilot knows the river. Fourteen of the islands and a small area of mainland constitute the Thousand Islands National Park. Most of the other islands are in private ownership, a favorite resort for residents of Canada and the United States.

Canadian National Parks Bureau

A VIEW FROM THE BASE OF FRENCH MOUNTAIN

Meandering along the rugged coastline of Nova Scotia, Cabot Trail leads the traveler to many vistas of indescribable charm. Cabot Trail is within Cape Breton Highlands National Park, which occupies the northern part of Cape Breton Island. The coastline here has been likened to the Highlands of Scotland and many people from Scotland have settled in the vicinity.

portance in Canada. This work was commenced in 1919, with a view to preserving and maintaining as a national heritage the sites and relics associated with stirring events in Canadian history. In this work the Bureau is assisted by the Historic Sites and Monuments Board of Canada, an honorary body whose members reside in various parts of the country and are historians of outstanding reputation. Since the inception of the work, more than three hundred sites have been judged to be of sufficient national importance to warrant marking by suitable memorials.

Further extensions to the National Parks system were made in 1941 when seven areas, previously acquired and administered as historic sites, were designated National Historic Parks—Louisbourg Fortress and Port Royal National Historic Park, Nova Scotia; Fort Chambly and Lennox, Quebec; Forts Wellington and Malden, Ontario; and Fort Prince of Wales, Manitoba. Fort Langley in British Columbia is a historic site. Preservation or restoration work has been carried out at all of these points, and at some of the old forts, historical museums have been constructed or arranged for, to house exhibits or relics.

Not all memorials, however, are dedicated to commemorate warlike episodes. At Gaspé, Quebec, a huge granite cross thirty feet high marks the landing place of Jacques Cartier in 1534. At Charlottetown, Prince Edward Island, a bronze tablet commemorates the laying of the first submarine telegraph cable in America in 1852, and at Halifax, Nova Scotia, a tablet calls attention to the establishment in 1752 of the first newspaper in Canada. Hardly a section of Canada is without a point of historic interest, and it is hoped that eventually every site of national importance and interest in the Dominion will be preserved from oblivion and become an object of the nation's care.

In addition to the National Parks and Historic Sites which we have mentioned,

UNSPOILED WILDERNESS LANDS

most of the Provinces have established Provincial Parks for the pleasure and the health of the people. Some of these are larger than many of the National Parks, and compare favorably in beauty and interest with some of the parks established by the Dominion.

Quebec has set aside two large areas. The Laurentides, in the Laurentian Mountains, lies between Quebec City and Lake Saint John. It is reached by a good motor road from Quebec which passes through the park, and goes on to encircle Lake Saint John. The park contains 4,000 square miles of forests, lakes and streams, and is much used as a vacation resort by the people of the cities and towns. Camps and camp-sites have been provided by the Provincial Government. The number of visitors is sure to increase as more roads are built.

Montagne Tremblant Park, also in the Laurentians, eighty miles by rail north of Montreal, is a famous resort for both winter and summer tourists. There are forty miles of ski trails through extensive forests and countless lakes and streams well stocked with trout.

Ontario has six Provincial Parks, two of them large. The most famous is Algonquin Provincial Park, which includes 2,740 square miles. It boasts more than 1,200 lakes, generally connected by streams of varying size. It lies between the Ottawa River and Georgian Bay. Campers, canoeists and fishermen find it delightful. It is visited by thousands from Canada and the United States who

Canadian National Parks Bureau

GATEWAY TO CAPE BRETON HIGHLANDS NATIONAL PARK

The harbor at South Ingonish offers shelter to many fishing boats such as these. Sword-fishing is the principal industry of the people and some of them even venture into the Atlantic Ocean to catch the big fish with a harpoon. South Ingonish forms the eastern entrance to Cape Breton Highlands National Park where the traveler finds much of interest.

UNSPOILED WILDERNESS LANDS

Canadian National Parks Bureau
MADELEINE DE VERCHÈRES MEMORIAL
Overlooking the St. Lawrence River at Verchères, Quebec, is this fine monument erected in memory of the French Canadian heroine, Madeleine de Verchères.

wish to forget that there is any such thing as a city. It is a wild-life sanctuary, and the possession of firearms is forbidden. For those who wish more comfort than camping permits, there are several hotels. The park can be reached either by railroad or by motor. The headquarters of the administration are on Cache Lake. This park was established many years ago, and many of the first visitors return year after year.

Quetico Provincial Park is located on the International Boundary line about a hundred miles west of Fort William. The area is 1,860 square miles of virgin forest through which roam moose, deer, bear, and many other animals. Hunting is absolutely forbidden. There are many beautiful lakes, connected by rivers or streams, which teem with fish of many sorts. Canoeists and fishermen find the region a paradise. The park may be reached by the Canadian National Railways, but there are no motor roads as yet.

Rondeau Provincial Park is a small area of only eight square miles, on Lake Erie, south of Blenheim. It contains specimens of nearly every tree which grows naturally in southern Ontario. Deer, wild turkeys, pheasants, and beaver are found. Fishing and camping are permitted under restrictions.

An interesting park in Manitoba is the International Peace Garden, south of Boissevain. The project was urged by a group of gardeners, horticulturists in Toronto in 1929, and the Garden was dedicated in 1932. It contains 1,800 acres, half in the Province and half in North Dakota, and is almost equidistant from the Atlantic and the Pacific Oceans. It was established to commemorate the friendly relations which have existed between Canada and the United States for over a century. The Garden is being developed by private subscription and it is intended to include all flowers and shrubs which grow in the temperate zone. The Garden is easily reached by the motor highways of the Province and of North Dakota.

Saskatchewan has established nine Provincial Parks, none of them very large. Cypress Hills, with an area of 17 square miles is one of the highest elevations in the Province. There are many trout streams, and boating, bathing and golf may also be enjoyed. Duck Mountain Park, near Kamsack, lies against the Manitoba boundary, and is 81 square miles in extent. Several lakes afford good boating and bathing. Wild life is abundant, including many species of waterfowl. There are bungalow camps, cabins and camp sites.

UNSPOILED WILDERNESS LANDS

Greenwater Lake Park, thirty-five square miles in area, provides boating, bathing and fishing, and camp sites have been provided. Good Spirit Lake Park, one of the smallest, contains only six square miles. It is located near Canora, and provides good fishing and bathing. There is a tourist camp.

Moose Mountain Park, not far from Carlyle, contains 154 square miles. Much has been done for the comfort of visitors, and the accommodations are more developed than in some of the other parks. There are hotels, cabins and camps. Fishing, boating, bathing and golf are the recreations. The largest of all the Provincial Parks in Saskatchewan is Nipawin, 252 square miles, which lies to the east of Prince Albert National Park. So far the park is undeveloped.

Alberta has six National Parks and has also set aside twenty-five areas for Provincial Parks, though little development work has been done.

British Columbia has already developed fifteen Provincial Parks and has reserved many other areas. One of the newest and by far the largest of these parks is Tweedsmuir, 5,400 square miles in area. It is also the farthest north, lying about two-thirds the distance between the city of Vancouver and the border of Alaska. Forest, lake, mountain and stream lie within its boundaries affording a variety of scenery. The Park abounds with wild life; its waters provide excellent fishing.

Garibaldi Provincial Park, northeast of Vancouver, is in the heart of the Coast Mountains. It contains 973 square miles of mountain scenery with lakes, waterfalls and glaciers. There is an abundance of wild life, which is carefully protected. Many of the most beautiful views can be reached only by trail.

Canadian National Parks Bureau

THE MUSEUM AT FORT ANNE, NOVA SCOTIA

In 1605, Champlain built a fort in what we now know as Nova Scotia. It was later moved to the present site, and during the contest between French and English, changed hands six times. The building you see through the gateway is the officers' quarters built by the Duke of Kent, the father of Queen Victoria, in 1797. It is now a museum.

UNSPOILED WILDERNESS LANDS

Kokanee Glacier Park, just west of Kootenay Lake, has an area of a hundred square miles. There are several high peaks, but the Kokanee glacier is the outstanding feature. Mount Assiniboine Park lies next to Banff National Park, and contains only twenty square miles, but it is studded with beautiful glacier-fed lakes, and the fishing in the lakes and streams is good. Naturally most of the park can be reached only by trail. A bungalow camp near Lake Magog is open in the summer.

Mount Robson Park is really an addition to Jasper National Park and contains 803 square miles. The park area is "a sea of peaks divided by the valley of the Fraser River." Mount Robson rises to a height of 12,972 feet. The Fraser River has its source among the glaciers in the park. The park is crossed by the main line of the Canadian National Railways. There are cabin accommodations.

Strathcona Provincial Park, 828 square miles in extent, is on Vancouver Island. There is alpine scenery of great beauty besides primeval forests, glaciers, mountain lakes and crystal streams. So far no tourist accommodations have been developed. The many areas set aside for additional provincial parks are all small. They range in size from Sooke Mountain, 1,446 acres to Quesnel, only four acres in extent. All of these areas are attractive for one reason or another.

Doubtless other areas will be established as parks both by the Dominion and the Provinces. More and more people are realizing that beauty and recreation are necessary for the well-being of all the people. Cities after all are artificial and it is good for us to see at frequent intervals what Nature has to offer, before the sights have been spoiled by the hands of man. No country has been more favored by Nature than Canada and it is well that the governments are striving to preserve the wonderful heritage.

An interesting fact is the number of residents of the United States who visit the parks and the historic sites of Canada. Banff, Jasper, Waterton, Algonquin and others are almost as well known in the United States as in Canada. Every year thousands cross the border by railway, or by motor, to spend the whole or a part of their vacations in Canada. Many of these visit one or more of the parks, or else see some of the historic sites of the Dominion. On any summer day one may see cars with license plates from many states in any of the more important parks, or else at some historic site. The number of visitors from the United States is likely to increase.

Canadian National Parks Bureau

IN MEMORY OF JACQUES CARTIER

This tall granite cross overlooking the St. Lawrence River was erected in 1934 to commemorate the landing of Jacques Cartier at Gaspé just four hundred years earlier.

CANADA'S NATIONAL PARKS: FACTS AND FIGURES

RECREATIONAL PARKS
(Date established, area in square miles and location)

Banff (1885), 2,564, in western Alberta; Cape Breton Highlands (1936), 390, in northern part of Cape Breton Island; Georgian Bay Islands (1929), 5.4, in Georgian Bay north of Midland, Ontario; Fundy (1948), 80, on Bay of Fundy between Moncton and St. John, New Brunswick; Gatineau, 50, in southwestern Quebec; Glacier (1886), 521, in southeastern British Columbia; Jasper (1907), 4,200, in western Alberta; Kootenay (1920), 543, in southeastern British Columbia; Mount Revelstoke (1914), 100, in southeastern British Columbia; Point Pelee (1918), 6, on the Lake Erie shore of Ontario; Prince Albert (1927), 1,496, in central Saskatchewan; Prince Edward Island (1937), 7, on north shore of Prince Edward Island; Riding Mountain (1929), 1,148, in southwestern Manitoba; St. Lawrence Islands (1914), 189.4 (acres), in St. Lawrence River between Morrisburg and Kingston, Ontario; Waterton Lakes (1895), 204, in southern Alberta; and Yoho (1886), 507, in eastern British Columbia.

WILD ANIMAL PARKS

Buffalo (1908), 197.5, near Wainwright, Alberta; Elk Island (1913), 75, near Edmonton in central Alberta; Nemiskam, 8.5, near Foremost, Alberta; and Wood Buffalo (1922), 17,300, partly in the Northwest Territories and partly in Alberta.

HISTORIC PARKS
(Area in acres)

Fort Anne (1917), 31, at Annapolis Royal in Nova Scotia; Fort Battleford, 36.7, near North Battleford, Saskatchewan; Fort Beauséjour (1926), 81, near Sackville in New Brunswick; Fort Chambly (1941), 2.5, in Chambly, Quebec; Lower Fort Garry, 13, near Winnipeg, Manitoba; Fort Lennox (1941), 210, near St. Johns, Quebec; Fortress of Louisbourg (1941), 340, on Cape Breton Island, Nova Scotia; Fort Malden (1941), 5, in Amherstburg, Ontario; Fort Prince of Wales (1941), 50, near Churchill in Manitoba; Fort Royal (1941), 17, in Lower Granville, Nova Scotia; and Fort Wellington (1941), 8.5, in Prescott, Ontario.

HISTORIC SITES

The Canadian Government has also restored and marked more than 500 places of national historic interest.

INTERNATIONAL PEACE GARDEN

This park is maintained by Canada and the United States and straddles the international boundary, 1,451 acres being in Manitoba and 888 acres in North Dakota.

CANADA'S PROVINCIAL PARKS: FACTS AND FIGURES
(Date established, area in acres and location)

Alberta

Bad Lands Reserve (1934), 1,800, north of Drumheller; Crimson Lake (1948), 900, near Rocky Mountain House; Elkwater Lake (1947), 378, Cypress Hills; Ghost River (1930), 536, west of Calgary; Kinbrook Island (1949), 90, Lake Newell; Red Lodge (1948), 45, west of Bowden; Saskatoon Island (1932), 250, west of Grande Prairie; Saskatoon Mt. Reserve (1930), 3,000, in Grande Prairie District; Writing-on-Stone Reserve (1930), 796, northeast of Coutts.

British Columbia

Beatton (1934), 770, near Fort St. John; Chasm (1940), 315, Clinton; Clearwater (1938), 260, near Hadly; Crescent Beach (1938), 237, near U. S. border; Cultus Lake (1948), 950, Chilliwack; Darke Lake (1943), 5,472, South Okanagan, Elk Falls (1940), 2,558, Vancouver Island; Garibaldi (1927), 612,615, near Haney-Squamish; Hamber (1941), 2,431,960, adjoins Banff and Jasper; Kokanee (1922), 64,000, near Nelson; McMillan (1944), 337, near Cameron Lake; Manning (1941), 181,760, near U. S. border; Mt. Assiniboine (1922), 12,800, south of Banff; Mount Bruce (1938), 480, Salt Spring Island; Mt. Maxwell (1938), 492, on same island; Mt. Robson (1913), 513,920, near Jasper Park; Mt. Seymour (1936), 9,156, near U. S. border; Nakusp Hot Spring (1925), 127, near Arrow Lakes; Peace Arch (1939), 16, international boundary; Silver Star (1940), 21,888, Vernon; Sir Alexander Mackenzie (1926), 13, Ocean Falls; Sooke Mt. (1928), 1,446, Vancouver Island; Stamp Falls (1940), 424, Vancouver Island; Strathcona (1911), 529,920, center of Vancouver Island; Tow Hill (1948), 480, Queen Charlotte Island; Tweedsmuir (1938), 3,456,000, near Burns Lake; Wells Grey (1939), 1,164,960, north of Kamloops.

Newfoundland

Serpentine (1939), 26,800, on west coast.

Ontario

Algonquin (1893), 1,754,240, in the southeast; Ipperwash Beach (1937), 109, Lambton County; Lake Superior (1944), 345,600, in the District of Algoma; Quetico (1913), 1,190,400, near international boundary between Port Arthur and Fort Frances; Rondeau (1894), 5,120, Kent County; Sibley (1944), 40,320, in Thunder Bay District.

Quebec

Chibougamau Fish and Game Reserve (1946), 1,088,000, west of Lake St. John; Gaspesian (1937), 320,000, Gaspé Peninsula; Lac Kippewa Fish and Game Reserve, 640,000 acres; Laurentides (1895), 2,373,120, north of Quebec City; Mt. Orford (1938), 9,970, west of Sherbrooke; La Verendrye (1939), 1,732,000, in the west; Trembling Mt. (1895), 770,500, north of Montreal.

Saskatchewan

Cypress Hills (1932), 10,880, near international boundary; Duck Mt. (1932), 51,840, northeast of Kamsack; Good Spirit Lake (1932), 3,827, west of Canora; Greenwater Lake (1932), 22,240, north of Kelvington; Lac La Ronge (1939), 729,600, north of Prince Albert; Moose Mt. (1932), 98,560, north of Carlyle; Nipawin (1934), 161,280, northwest of Nipawin.

Color Plates in Volume VI

ALASKA:

	PAGE
Sitka	311

CANADA:

	PAGE
Girl Fishing	50
Fishing, Restigouche River	51
Black Harbor, New Brunswick	54
Peggy's Point, Nova Scotia	55
Making Hay, Nova Scotia	58
Gaspereau Valley	59
Margaree Valley	62
Prince Edward Island	63
Château de Ramezay	66
St. Louis Gate, Quebec	67
Sous-le-Cap Street, Quebec	70
Tadoussac Landing	71
Quebec Orchards	74
Boys with a Dog Cart	75
Montmorency Falls, Quebec	78
Cape Trinity, Quebec	79
Cross Lake, Cobalt	82
Peach Orchards, Ontario	82
Channel, Lake of Bays	83
Sunnyside Beach	83
Young Cattle, Ottawa	86
Lumber Jam, Ontario	86
Vineyards in Ontario	87
Ottawa River	87
Main Channel	90
Sault Ste. Marie	91
Governmental Driveway	94
Peach Orchards in Bloom	95
Takkakaw Falls	98
Main Channel	99
Massive Mountain Range	102
Moraine Lake, Alberta	103
Prince Albert Park	106
The Selkirk Mountains	107
Bow Falls	110
Emerald Lake	111

HAWAII:

	PAGE
Natural Bridge	306
Waikiki Beach	307
From Nuuanu Pali	310

NORTH AMERICAN INDIANS:

	PAGE
Navaho Women	18
Thunderbirds Blanket	19
Sand Painting	22
Zuñi Craftsman	23
Iroquois Mask	26
Seminole Robe	27

NORTH AMERICAN INDIANS (*continued*):

	PAGE
Sioux Costume	30
Totem Poles	31
Hopi Indian	274
Walapai Woman	275
Making Pottery	278
Apache Brave	279
Pueblo Chief	282
Rain Dance of the Zuñis	283

UNITED STATES:

	PAGE
Independence Hall	186
Mount Vernon	187
Washington Monument	190
Pan-American Union	191
Making Maple Sugar	210
Mulberry Street, New York	211
Devil's Pulpit, Maine	214
Niagara Falls	215
Delaware Water Gap	218
White Horse Ledge	219
Chew House	222
Johnston Gate, Harvard	223
The Swannanoa	226
Lookout Mountain	227
Harper's Ferry	230
Texas Plains	231
Courtyard, New Orleans	234
Christ Church, Alexandria	235
Bruton Parish House	235
Bay Saint Louis	238
Magnolia Gardens	239
Arch Rock, Mackinac Island	258
Fort Snelling	259
Tavern, Lafayette Ohio	262
Cedar River	263
White River, Ozark Plateau	266
Black Hills	267
Rocks near Camp Douglas	270
Horseshoe Fall, Illinois	271
San Luis Rey	286
A Giant Cactus	287
Mount of the Holy Cross	298
Canyons of the Colorado River	299
Multnomah Falls	302
California Gardens	303
Grand Canyon	386
Emerald Spring	387
Glacier National Park	390
Crater Lake	391
Yosemite Falls	394
Grand Canyon	395
Zion National Park	398
Bryce Canyon	399

ARCTIC OCEAN
Ellesmere I.
GREENLAND
Baffin Bay
Bank I.
Victoria I.
Baffin I.
ASIA
Bering Strait
Yukon River
Mackenzie River
Great Bear Lake
Great Slave Lake
Hudson Bay
Davis Strait
Bering Sea
Aleutian Islands
ROCKY MTS.
R. Nelson
L. Winnipeg
Vancouver Island
NORTH AMERICA
Great Lakes
St. Lawrence R.
Newfoundland
Columbia R.
Missouri R.
APPALACHIAN MTS.
Azores
Colo. R.
Rio Grande
Miss. R.
ATLANTIC
Bermuda I.
Gulf of Mexico
Bahama Is.
West Indies
Cuba
Hawaii Islands
CENTRAL AMERICA
Caribbean Sea
OCEAN
PACIFIC
Panama Canal
Orinoco R.
Gilbert Is.
EQUATOR
Amazon River
Phoenix Is.
Solomon Is.
Marquesas Is.
ANDES
SOUTH AMERICA
Madeira R.
R. Sao Francisco
Society Is.
New Hebrides
OCEAN
Fiji Is.
New Caledonia
Tuamotu Archipelago
MTS.
Parana R.
NEW ZEALAND
Falkland Islands
Strait of Magellan
Cape Horn
Drake Strait
Weddell Sea
ANTARCTIC
CIRCLE